GeoPlanet: Earth and Planetary Sciences

For further volumes:
http://www.springer.com/series/8821

Roman Teisseyre · Maria Teisseyre-Jeleńska

Asymmetric Continuum

Extreme Processes in Solids and Fluids

Springer

Roman Teisseyre
Maria Teisseyre-Jeleńska
Institute of Geophysics
Polish Academy of Sciences
Warsaw
Poland

The GeoPlanet: Earth and Planetary Sciences Book Series is in part a continuation of Monographic Volumes of Publications of the Institute of Geophysics, Polish Academy of Sciences, the journal published since 1962 (http://pub.igf.edu.pl/index.php).

ISSN 2190-5193 ISSN 2190-5207 (electronic)
ISBN 978-3-642-31859-7 ISBN 978-3-642-31860-3 (eBook)
DOI 10.1007/978-3-642-31860-3
Springer Heidelberg New York Dordrecht London

Library of Congress Control Number: 2013939064

Printed on acid-free paper

Springer is part of Springer Science+Business Media (www.springer.com)

Series Editors

Geophysics: Paweł Rowiński
Editor-in-Chief
Institute of Geophysics
Polish Academy of Sciences
ul. Ks. Janusza 64
01-452 Warszawa, Poland
p.rowinski@igf.edu.pi

Space Sciences: Marek Banaszkiewicz
Space Research Centre
Polish Academy of Sciences
ul. Bartycka 18A
00-716 Warszawa, Poland

Oceanology: Janusz Pempkowiak
Institute of Oceanology
Polish Academy of Sciences
Powstańców Warszawy 55
81-712 Sopot, Poland

Geology: Marek Lewandowski
Institute of Geological Sciences
Polish Academy of Sciences
ul. Twarda 51/55
00-818 Warszawa, Poland

Managing Editor

Anna Dziembowska
Institute of Geophysics, Polish Academy of Sciences

Advisory Board

Robert Anczkiewicz
Research Centre in Kraków
Institute of Geological Sciences
Kraków, Poland

Aleksander Brzeziński
Space Research Centre
Polish Academy of Sciences
Warszawa, Poland

Javier Cuadros
Department of Mineralogy
Natural History Museum
London, UK

Jerzy Dera
Institute of Oceanology
Polish Academy of Sciences
Sopot, Poland

Evgeni Fedorovich
School of Meteorology
University of Oklahoma
Norman, USA

Wolfgang Franke
Geologisch-Paläntologisches Institut
Johann Wolfgang Goethe-Universität
Frankfurt/Main, Germany

Bertrand Fritz
Ecole et Observatoire des
Sciences de la Terre
Laboratoire d'Hydrologie
et de Géochimie de Strasbourg
Université de Strasbourg et CNRS
Strasbourg, France

Truls Johannessen
Geophysical Institute
University of Bergen
Bergen, Norway

Michael A. Kaminski
Department of Earth Sciences
University College London
London, UK

Andrzej Kijko
Aon Benfield
Natural Hazards Research Centre
University of Pretoria
Pretoria, South Africa

Francois Leblanc
Laboratoire Atmospheres, Milieux
Observations Spatiales – CNRS/IPSL
Paris, France

Kon-Kee Liu
Institute of Hydrological
and Oceanic Sciences
National Central University Jhongli
Jhongli, Taiwan

Teresa Madeyska
Research Centre in Warsaw
Institute of Geological Sciences
Warszawa, Poland

Stanisław Massel
Institute of Oceanology
Polish Academy of Sciences
Sopot, Polska

Antonio Meloni
Instituto Nazionale di Geofisica
Rome, Italy

Evangelos Papathanassiou
Hellenic Centre for Marine Research
Anavissos, Greece

Kaja Pietsch
AGH University of Science and
Technology
Kraków, Poland

Dušan Plašienka
Prírodovedecká fakulta, UK
Univerzita Komenskèho
Bratislava, Slovakia

Barbara Popielawska
Space Research Centre
Polish Academy of Sciences
Warszawa, Poland

Tilman Spohn
Deutsches Zentrum für Luft-
und Raumfahrt
in der Helmholtz Gemeinschaft
Institut für Planetenforschung
Berlin, Germany

Krzysztof Stasiewicz
Swedish Institute of Space Physics
Uppsala, Sweden

Roman Teisseyre
Earth's Interior Dynamics Lab
Institute of Geophysics
Polish Academy of Sciences
Warszawa, Poland

Jacek Tronczynski
Laboratory of Biogeochemistry
of Organic Contaminants
IFREMER DCN_BE
Nantes, France

Steve Wallis
School of the Built Environment
Heriot-Watt University
Riccarton, Edinburgh
Scotland, UK

Wacław M. Zuberek
Department of Applied Geology
University of Silesia
Sosnowiec, Poland

Preface

The complexity of processes in the vortex motions in fluids and in a fracture source in solids has inspired us to reconsider a class of basic motions and deformations in the Asymmetric Continuum Theory. First of all, we consider the basic deformations in solids and molecular deformations in fluids and, moreover, the transport velocity in fluids and molecular transport in solids; we believe that these elements play an important role in the fracture processes. This approach gives grounds for considering the Asymmetric Continuum as a common basis for the continuum media with respective phase transitions. Thus, we consider the molecular deformations and transport velocities in fluids and will supplement the strain deformations in solids by molecular transport, important when the material becomes close to fracture.

In solids, the simple deformations include the axial nuclei, that is, point extension/compression, the shear nuclei and, when considering an asymmetric continuum, the rotation strains. Such rotation strains describe the oscillations of the main shear axes and their amplitudes. All these fields should be related to the independent motion equations. However, a separate problem concerns the displacements and a simple rotation which might be reduced even to a point dimension (Planck length). Recording of displacements, e.g., by means of seismometers, might prove their physical existence; however, we will show that such displacement records prove only the existence of displacement derivatives. The existence of displacement derivatives in solids allows us to define the molecular displacements, which might relate to the space and time derivatives. This approach will lead us to a new fracture criterion and permits to discuss some complex deformation processes. Here, an important role is played by the defect distributions of opposite signs, leading to the induced opposite-sense strains, and fracture processes.

In fluids, the transport displacement velocities describe the motion processes, while the molecular strains are related to the time rates of strains or deformations. Here, a special role is played by vorticity. The vortex motions are usually related to rotation of transport motion; however, our asymmetric approach to the continuum theory allows us to define the vortex motions by means of the rotational transport expressed in a cylindrical system. In some publications we have already shown that this new definition opens a way leading to more complicated problems, like those related to turbulence phenomena, including an influence of the gas–liquid phase transitions at breaks of some molecular bonds inside some part of a continuum.

Further, we show how this new approach to the Asymmetric Continuum, as described by a joint theory, permits also to discuss and introduce new elements into the theory of defects. The defect densities form the induced strains and effectively change the internal stress system; this approach explains many important problems related to the fracture processes. We include also some thermodynamical problems, especially those related to the fracture processes. Some quantum theory analogies help us to better understand the extreme processes in fluids and solids. We also consider and try to explain the extreme processes, like fractures in solids and turbulence motions in fluids.

Further, we explain how this new approach to the Asymmetric Continuum, as described by a joint theory, permits also to discuss the complex interaction processes, especially those related to the electric and magnetic fields including the mutual influences between these fields and the stresses. The interaction processes and some thermodynamical problems and quantum theory analogies help us to understand the extreme processes in fluids and solids. A final problem is the release–rebound theory and the interaction processes and the wave propagations presented in an invariant four-dimensional form.

Acknowledgments

The authors would like to express their sincere acknowledgments to Prof. P. Varotsos for his valuable remarks.

We are also very thankful to Prof. E. Dologlou. Important remarks and comments have been provided by Dr. K. P. Teisseyre. We express special thanks to Managing and Language Editor of the Series, Anna Dziembowska, for her very effective help and important corrections. Most of the figures were prepared by Dr. Marek Górski.

Contents

Chapter 1
Introduction: Independent Strain and Transport Motions

1.1 Introduction

The complexity of processes in the vortex motions in fluids and in a fracture source (e.g., seismic source) in solids, has inspired us to reconsider a class of basic motions and deformations in the Asymmetric Continuum Theory. First of all, we consider the basic deformations in solids and molecular deformations in fluids and, moreover, the transport velocity in fluids and molecular transport in solids. We believe that this new element, molecular transport in solids, plays an important role in the fracture processes. Our approach gives grounds for considering the Asymmetric Continuum as a common basis for the continuum media, solids and fluids. In solids, the simple deformations include the axial nuclei, that is, point extension/compression, the shear nuclei and, when considering an asymmetric continuum, the rotation strains. Such strain rotations describe the oscillations of the main shear axes and their amplitudes. All these fields should be related to the independent motion equations. However, a separate problem is related to displacements and a simple rotation which might be reduced even to a point dimension (Planck length). Recording of the displacements, e.g., by means of seismometers, might prove their physical existence; however, we will demonstrate that such displacement records prove only the existence of displacement derivatives. The existence of displacement derivatives in solids allows us to define the molecular displacements, which might relate to the space and time derivatives. This approach leads us to a new fracture criterion and permits to discuss some complex deformation processes. Here, an important role is played by the defect distributions of opposite signs, leading to the induced opposite-sense strains and fracture processes. In solids without any fractures, the displacement motions cannot exist: here the main role is due to deformation fields, the axial, shear and rotation strains. These strain fields may be quite independent; their description can be presented with the help of a different reference displacement. The reference displacement means the displacements which do not exist in reality, being only introduced to describe these strain fields mathematically. Of course, the strain

R. Teisseyre and M. Teisseyre-Jeleńska, *Asymmetric Continuum*,
GeoPlanet: Earth and Planetary Sciences, DOI: 10.1007/978-3-642-31860-3_1,

waves can be recorded by seismometers and such records are again presented as displacements; we explain this paradox.

1.2 Strain and Transport Motions

The Asymmetric Continuum Theory related to solids (Teisseyre 2008a, b, 2009, 2011) has been constructed to avoid the numerous insufficiencies of the classic symmetric theory and some phenomena found by seismological observations. At the end of twentieth century, the records of rotation motions have been obtained by means of the Sagnac measurement system (cf., Lee et al. 2009).

Between the mathematics of the continuum theory and the physics of wave displacements there appears a controversy; mathematically the wave displacements should not exist, while in reality they appear only in the range of a molecular structure without any material breaks; the related material constants are estimated from the wave records. To maintain the consistency of mathematical description we will define molecular displacements in solids, similarly to the molecular strains used in fluids; in this way the molecular fields will help us to present the joint, both for solids and fluids, Asymmetric Continuum Theory. In the frame of such a theory we also explain the role of rotational motions, including the molecular rotation strains.

Thus, while constructing the Asymmetric Continuum Theory related to both the solid and fluid states we will reject any initial theoretical limitations.

We will also reconsider the independent motions and deformations in the frame of the considered continuum. We hope that such an approach will explain some extreme processes, e.g., in the seismic sources in solids, and the turbulence phenomena in fluids.

In solids, we usually have in mind the displacements and also shear and axial strains; however, in a general approach we might consider both the displacements and the rotations as point motions (defined by the Planck length, ca 10^{-35} m) but instead we will introduce the rotation strains complementary to shear strains.

In solids, a separate problem relates to the displacements; we will explain why their recording, e.g., by means of seismometers, prove only the existence of molecular displacements, or displacement derivatives, but not the displacements themselves. Instead we will define the molecular displacements and new fracture criterion including the defect distributions and induced strains.

In fluids, the transport velocities and molecular strains describe the motion processes. Usually the vortex motions are defined by means of the rotational transport; this approach leads us to more complicated problems, like the turbulence phenomena.

However, we should remember that many attempts have been made to improve the classic elasticity. First, we may note the problem related to rotations; these motions are not included in a consistent way in the classical elasticity. Since the end of the nineteenth century, many attempts have been undertaken to construct a more adequate and powerful theory of continua. A first attempt to include rotations into the continuum theory is due to Voigt in 1887, while a complete theory, with the displacement vector and rotation vector, was proposed by the Cosserat brothers in 1896; (see Cosserat and Cosserat 1909). Much later, a very powerful approach to a continuum has been provided by the elastic micropolar and micromorphic theories developed by Eringen and his coworkers (see, e.g., Eringen and Suhubi 1964) and Mindlin (1965). A simpler theory, the Asymmetric Elasticity, proposed by Nowacki (1986), is also worth mentioning. For the advanced review of the micromorphic theory for solids and fluids we may mention two important monographs by Eringen: "Microcontinuum Field Theories I, Foundations and Solids" (Eringen 1999) and "Microcontinuum Field Theories II, Fluent Media" (Eringen 2001). These micropolar and micromorphic theories present a very powerful tool to describe many complicated material problems (see: Jones 1973). These theories seem, however, quite complicated and difficult in a common use, partly due to a number of additional constants not well determined by experiments. For some trials of applying these theories into seismology, see Teisseyre (1973, 1974).

Another approach to generalize the classic theory is due to Kröner's papers; in that theory the elastic fields are given as the difference between the total fields and the self-fields represented as a continuous distribution of internal nuclei; e.g., Kröner (1981).

Our intention is to remain in the linear domain; however, we may mention some problems solved with the help of nonlinear relations, e.g., the theory of the soliton waves, important in ocean dynamics, but also recently discussed in seismology (Mikhailov and Nikolaevskii 2000; Bykov 2006, 2008). Another way was to use the Riemannian or non-Riemannian space geometry (e.g., Teisseyre 1995).

Of course, the Classical Elasticity almost perfectly describes the small deformations, but there remain many well-recognized insufficiencies in it, e.g., the angular motions are not coherently incorporated. In this classic approach the balance of angular momentum holds only for the symmetric stresses, and the angular motions can be introduced only artificially, assuming some characteristic length element and a reference rotation point.

The main inadequacies and unsolved problems faced in the classical theories are as follows:

- In the classical elastic theory the balance of angular momentum holds only for the symmetric stresses, while the angular motions can be only introduced in an artificial way with the help of a characteristic length element and a reference rotation point. The basic strain motions (axial, shear and rotation strains) do not present any independent deformations. The point is that the independently released strain fields should remain independent, as in particular a balance of angular momentum which either does not exist or is introduced in an artificial way.
- Fracture pattern reveals usually an asymmetric pattern with the main slip plane as fracture deformations with the granulation and fragmentation processes which include the rotation motions;

- While searching for the fault-slip solutions we should use the friction constitutive laws introduced additionally, in accordance with the experimental data; these elastodynamic solutions properly describe the slip propagation along the fault, including friction effects and related seismic radiation.
- Advanced deformations, like plastic flow, can only be accomplished by changing the constitutive laws, while more serious problems appear when we want to describe the granulation and fragmentation of material, which needs an inclusion of rotation processes in the frame of coherent theory with asymmetric motions.
- Geometry of earthquake fracture usually reveals an asymmetric pattern and some premonitory processes develop in an asymmetric way.
- Solution for an edge dislocation presents some asymmetry in the strain components in the plane perpendicular to the dislocation line (wedge line); for a continuous distribution of dislocations this fact leads to confrontation between the symmetry of shear strains and asymmetry of stresses.
- Direct differential relation between the density of edge dislocations and stresses cannot be found in the frame of the classic symmetric continuum; the edge dislocations present the asymmetric strains. A relation between the density of edge dislocations and stresses can be found in the frame of the asymmetric theory, while it does not exist in the classic elasticity.

Similar considerations can be presented for fluids; the molecular strains produce symmetric and anti-symmetric molecular deformations (Teisseyre 2009). However, we meet there additional complications due to the non-laminar and turbulent motions and the density variation, mainly in gas state.

We will present the new approach related to these problems, although we should underline that similar efforts have been already undertaken in terms of the micropolar and micromorphic theories (Eringen 2001).

In another approach, the Kröner method (Kröner 1981), the physically significant elastic fields, $S_{ks}, E_{ks}, \omega_{ks}$, are given by the differences between the total fields, $S^T_{ks}, E^T_{ks}, \omega^T_{ks}$, related directly to the displacement derivatives, and the self fields, $S^S_{ks}, E^S_{ks}, \omega^S_{ks}$ (related to internal interaction nuclei): $S_{ks} = S^T_{ks} - S^S_{ks}$, $E_{ks} = E^T_{ks} - E^S_{ks}, \omega_{ks} = \omega^T_{ks} - \omega^S_{ks}$. Only the total field preserves the usual symmetry properties: elastic and self fields may be asymmetric. A comparison of our approach and that used in the Kröner method was given by Teissyere (2008); we recall here only that the interaction fields in the Kröner theory enter through the self-nuclei whose fields appear in the self-stress, self-strain and self-rotation fields. In the Kröner theory, the elastic fields represent the physical fields; the total field preserves the usual symmetry properties, while the elastic and self fields may be asymmetric.

Starting to present a new basis for the continuum theory we should consider the fundamental point deformations as discussed by Teisseyre and Górski (2009a, b). Considering the point motions first we have in mind the displacements and rotations. The problem of rotation motion and its recording appeared very early: the effects of the Lisbon (1755) and Calabria (1783) earthquakes. The founders of seismology: Lyell (1797–1875), Darwin (1809–1882), Mallet (1810–1881), and Humboldt (1769–1859), considered a counterpart of the rotation movements as an essential part of the motions induced by earthquakes. However, the majority of cases is related to the rotation effects on surface, where we should account for the geometry of the main inertia moments of an object and seismic acceleration. Of importance is also the position of the centre of adherence inside the ground (Ferrari 2006; Kozák 2006). However, there were also arguments against including the rotational motions into theory. Numerous attempts have been undertaken to record the rotation motions; the first instrument to record rotations was constructed by Filippo Cecchi in 1875, although this seismograph has never managed to record any traces of such waves. According to Gutenberg (1926) the rotations cannot propagate due to their immediate attenuation; however, as presented in our further discussions, a propagation of rotation waves is related to the interaction between the rotation strains and shear strains and in the classic theory the stresses are symmetric and thus the material strains should not include any rotation motions. In the Asymmetric Elasticity (Nowacki 1986) the postulate of the Central Symmetry argues against a direct use of rotations. This postulate means that an inversion of the coordinate system should not change any obtained solutions and in this way it eliminates the point rotations as the point rotations become oppositely oriented after a change of coordinate orientation. However, we question a need of using the point rotations and also the displacements in our Asymmetric Continuum Theory and therefore the Central Symmetry postulate is not needed. Of course, the rotation strains may exist in a modified Asymmetric Theory.

The recording of displacements, e.g., by means of seismometers, proves only the existence of displacement derivatives: the recorded displacement data refer to the total sum of all the deformations released. We believe that the true displacements relate to the slip motions along the faults and might be even observed at the faults situated at the surface. Yet the recorded displacements belong to the displacement derivatives related to a sum of the released strains and should be treated as the reference displacements belonging to different strains. It is even very difficult to imagine a release and, afterword, also a record of the very long displacement waves; the related displacement do not exist, but the deformations, that is, the displacement derivatives, really exist and the seismometers could record the displacement derivatives as an integrated effect related to an appropriate length element, Δx, of a very rigid seismometer platform, much more rigid than the soil layers.

Thus, the displacement:

$$u_k \approx \frac{\partial u_k}{\partial x_i} \Delta x_i, \quad k = i \pm 1 \tag{1.1a}$$

where Δx_i is the length of a seismograph platform (much more rigid than soil layers); the displacement, u_k, is perpendicular to it.

The rotation records could be estimated, e.g., in the following ways:

$$\omega = \frac{\partial \omega}{\partial \varphi} \Delta \varphi, \text{ or } \omega = \frac{1}{2\pi} \int_0^{2\pi} \frac{\partial \omega}{\partial \varphi} \frac{\Delta t}{T} d\varphi \tag{1.1b}$$

where the first equation relates to a measurement system of rotations by an array of seismometers, $\Delta\varphi \approx \frac{\Delta R}{V^S \Delta t}$, being an angular shift between the given seismometers and ΔR, the related distance (or more exactly its related projection), while V^S means the shear wave velocity given as equal to the strain rotation velocity.

The other expression,

$$\Delta T = 2\pi \frac{R}{c} \tag{1.1c}$$

relates to the Sagnac measurement system for the rotation velocity; here, R relates to a radius of the Sagnac circular recording system and c is the light velocity.

One of the aims of this book, in our opinion *extremely* important, is to substantiate the need for constructing the worldwide seismic network based on the strain and rotation recording systems. We should note that only in few seismic regions such recording systems already exist, which is not enough, as the strain waves can propagate over the whole world and influence the local activities.

One very important aspect is related to the long waves; we may observe an appearance of the very long waves, even of a length of some hundreds of kilometers, while the related displacement could be extremely small (see: Fig. 1.1).

There appears an evident controversy between the mathematics of the continuum theory for solids and the physics of wave displacements. In terms of mathematics, the related wave displacements should not exist in a continuum (under the assumption that there does not exist any fracture), while physically the displacement might appear when being very small, in a range of atomic or molecular structure, without any material breaks. We should note that the related material constants were estimated from the wave records. In our approach related to a continuum we assume that any displacements, not only those extremely small, can be treated only as the reference displacements. However, we may introduce the molecular displacements. In the frame of the continuum theory, we assume that there exist only the displacement derivatives. In a similar way, the point motions related to shears, or rotation amplitudes, are very small, but the shear strain, or rotation strain, related to the displacement derivatives, might be considerable.

The problem of fundamental point motions gains actuality due to the recent observations of rotation waves detectable by the very precise instruments based on the Sagnac effect and able to measure a very small spin, up to 10^{-9} rad/s. The

Fig. 1.1 Wavelength Λ, which can be very long, and the related displacements u, in a molecular range, will not be related to any material break

recorded rotation motions are not the simple point rotations, but do originate due to rotation strains. Our aim is to construct a new, relatively simple theoretical approach in which the strain moments would be replaced by the antisymmetric part of strains. When we introduce some displacements and rotations in the Asymmetric Continuum Theory, we will use them only as the reference motions. Moreover, these reference displacements could be shifted in phase when originated by the independent but correlated fracture processes.

Thus, in the next chapters related to our Asymmetric Continuum Theory, we define the molecular displacements and molecular transports in solids, and the molecular strains in fluids.

In this way a problem of fundamental point motions gains actuality; we may also note that, e.g., a single couple is not exactly a point source, but a double couple already becomes a point deformation with a shear nucleus given by the string-string deformation. Further developments of the Asymmetric Theory are due to Teisseyre and Boratyński (2003), and finally to Teisseyre (2008, 2009). Considering the fundamental motions as related to the strains, we believe that each of them should obey the adequate equations of motion. The recognized balance laws, including that for the moment of momentum, relate an earthquake source region to the premonitory and dynamic processes and the wave propagations. A general theoretical approach should include processes from a given state of continuum to other states governed by different constitutive laws; for example, we should include the links between the solid elastic continuum and that subjected to granulated and sand-like materials. We should also recognize the interactive processes between shear release and spin rebound and vice versa.

Thus, the fundamental point deformations lead to an axial strain (1 component) and to the shear strains (6 components) and the rotation strains (3 components); together we might deal with the 10 components. When remaining in a continuum, we should neglect the displacements and point rotations (Teisseyre and Gorski 2009; Teisseyre 2009, 2011). Theoretically, such motions might be recorded by the arrays of seismometers (Cochard et al. 2006). The axial deformation represents the compression/dilatation nuclei, related, e.g., to thermal anomaly, while the shear and rotation strains will be especially sensitive for a fracture process. We should also note that these deformations, including the axial motions, could be described by either the Riemannian curvature or, in more complicated cases, by the torsion tensor in a non-Riemannian space (Kleinert 1989).

There remains a problem of the scales of these motions generated in an earthquake source and also at sites on the ground surface where the similar constructions may give different responses to these wave arrived. According to the precise recording the strain motions, including the rotation ones, could be almost perfectly detectable by an array of seismographs and by the direct observation of rotations (e.g., by a Sagnac system, see: Igel et al. 2005, 2007; Cochard et al. 2006).

The observed rotational deformation on the Earth can be classified separately (Teisseyre et al. 2006) as follows:

- Megarotations: the ground tilts;
- Macrorotations: fragmentation processes at the fracturing especially under a compression load;
- Mesorotations, related to granulation processes and formation of the mylonite zones;
- Strain rotations in the wave motions directly interrelated with the shear strains and to fracture source processes;
- Microrotations in the internal friction processes, as well as caused by the microslip motions in dislocations and microcracks.

Our interest relates mainly to the strain rotations and our aim is to construct a new, relatively simple theoretical approach in which the stress moments would by replaced by the antisymmetric part of stresses, and the introduced shear and rotation motions could be shifted in phases as originated by the independent but correlated processes. For the antisymmetric part of stresses and related rotation we need to introduce a proper constitutive law, the Shimbo law (1975, 1995), related to the friction processes and rotation of grains. The equivalent constitutive laws have already been introduced in the micromorphic theories and in the Kröner approach to the continuum theory with the self-fields and related internal nuclei (Kröner 1981).

In our approach we include the strain rotations into a system of balance laws between stresses and strains with a possible phase shift between these fields.

Let us also note that in another way, when using the micromorphic theories, we must include the rotational motions in a quite complicated way. Yet in the classical elasticity the fault-slip solutions are introduced assuming some constitutive friction laws related to the rotation effects, as a description of granulation and material fragmentation needs to include rotation processes.

Moreover, we should note that a fracture geometry, e.g., due to an earthquake process, usually reveals an asymmetric pattern. The premonitory processes usually reveal an asymmetry; some of the previously mentioned insufficiencies of the classic theory found temporary solutions using the nonlinear relations, e.g., introducing the soliton waves, important in ocean dynamics and, for these reasons, we may find also many examples in seismological studies (e.g., Mikhailov and Nikolaevskii 2000; Bykov 2006, 2008); moreover, worth mentioning is the respective linear theory based on the Riemannian or non-Riemannian space geometry (Teisseyre 1995).

On the other hand, our efforts led us to a new approach assuming the existence of asymmetric strain fields; we assume that all the motions and deformations should relate to the separate balance laws, being either quite independent or mutually interrelated. We have two groups of relations, those for symmetric and antisymmetric fields: for the deformation we may write:

$$E_{kl} = \chi^E E_{(kl)} + \chi^\Omega E_{[kl]} \tag{1.2a}$$

where $|\chi^E| = 1, \quad \chi^\Omega = 0, \quad \text{or } \chi^E = 0, \quad |\chi^\Omega| = 1.$

For $E_{kl} = E_{(kl)}$ we obtain the classical elasticity with the *Central Symmetry* (Nowacki 1986), and for $E_{kl} = E_{[kl]}$ we will have a granular-crushed medium filled with rigid spherical grains with a friction interaction.

We arrive at the postulate opposite to the *Stable Position* (Nowacki 1986), where the applied torque load on a surface boundary will lead only to some angular deformation and the torque energy will be stored in it.

Another, more general case is given by the relations:

$$E_{kl} = \chi^E E_{(kl)} + \chi^\Omega E_{[kl]} \tag{1.2b}$$

where $\chi^E = \{\pm 1, \pm \mathrm{i}\}1$ and $\chi^\Omega = \{\pm 1, \pm \mathrm{i}\}$.

As we already stated, the problem of fundamental point motions gains actuality due to the recent observations of rotation waves. The rotation motions are not included in a classic theory. Our aim was to construct a new, relatively simple theoretical approach in which the stress moments would be replaced by the antisymmetric part of stresses, and the introduced displacement and rotation motions could be shifted in phase when originated by the independent but correlated processes. However, the existing theories either do not include all the point motions, or in an opposite case, assume a complete independence of these motions. Thus, we need a theory incorporating the separate balance relations and constitutive laws for all the point deformations, which will permit us to arrive at different solutions together with those presenting mutual interrelation between the fields as well as a possible phase shift between the motions. The existing technical facilities record the spin motion (Cochard et al. 2006), but we shall be aware that the recorded rotations come from different source processes related to the independent or mutually related translations and rotations in their origin sites. We may also note that, in the course of microfracturing under confining load, we are dealing with a significant rotation release process with the spin motion distinctly overpassing the displacement motions when observed at a very near field. For the strong near-ground motions, which include a tilting component, the rotation of displacements may be very important. Further on, we may even observe a magnification of horizontal rotation of displacements and an appearance of the rocking–tilting component of displacement rotation caused by the geometry of constructions, especially high buildings.

The shear and rotation strains act together; the oscillations of shear axes are schematically visualized in Fig. 1.2.

We conclude: the classic theory of elastic continuum has many basic inadequacies; it does not account for the following facts:

- When studying the elastic field of an edge dislocation we find some asymmetry in stress distribution; any direct differential relation between the density of dislocations and stresses cannot be found in a symmetric continuum theory.
- The earthquake slip distribution usually reveals its asymmetric pattern.
- The premonitory processes gradually lead from symmetric to antisymmetric pattern.

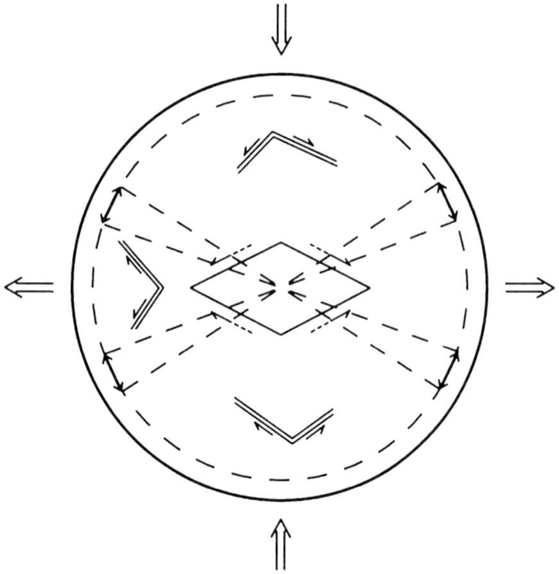

Fig. 1.2 Shear and rotational axes' oscillations; these motions, related to the inner microfracture in 4D, lead to a direct relation between these motions; we may notice rotations and transformation of an inner square into a rhomb

- The balance of angular momentum is introduced only artificially with the help of a length element and a reference rotation point.

In seismological research, when establishing the fault slip solutions, we rely on classical elasticity; however, we are obliged to introduce additionally the friction constitutive law based on the experimental data. A consistent continuum theory shall include the defect interactions and elastodynamic solution for the fault region and its friction properties included in an appropriate constitutive law.

Many trials have been made to improve the classical theory:

A first attempt is due to Voigt in 1887, although a complete theory including asymmetry of stress and strain is known as the Cosserat theory with the displacement vector and rotation vector (Cosserat and Cosserat 1909).

The micropolar and micromorphic theories were developed by Eringen and his co-workers (see, Eringen and Suhubi 1964; Eringen 1999).

The asymmetric theory of elasticity, including the couple-stresses, has been founded by Nowacki (1986).

The asymmetric theory, including asymmetric stresses, strains and rotations, and the related self-field (internal nuclei) according to the Kröner approach (Kröner 1981) has been proposed by Teisseyre and Boratyński (2003, 2006), and Teisseyre and Górski (2007); this theory refers to the Shimbo constitutive law joining the antisymmetric stresses and rotations (Shimbo 1975, 1995).

Very important arguments for the asymmetry of the basic strain fields come from the considerations on the circular squeeze strain deformation and wave.

Considering the shear load we may notice that the angular bond deformations produce centers with the squeeze angular strains, $E_{\varphi\varphi}$. Under compression load, the strains induced by defects produce centers with shears of opposite sense, and then some micro-breaks lead to the rotations and fragmentation processes, followed by the rebound slip motions retarded in phase. The shears create dynamic angular deformations leading to the bond breaks and slip propagation followed by the opposite-sense rebound rotations retarded in phase. On the contrary, the micro-fractures under compression lead to the induced shear motions of opposite sense. We may observe that the related fragmentations and granulations precede the slip rebounds, which appears retarded in phase.

However, we should take into our consideration that the strain waves in the classic theory are directly related to the displacement field, while in the asymmetric theory the strains could be treated as a completely independent physical field.

This interactive wave relation is rigorously supported by the invariant presentation of these fields in 4D (see: Chap. 12).

Of course, the different approaches should be applied to the rotation-strain waves propagation and to the slip and fragmentation motions forcing the material deterioration. Our basic assumption related to the propagation of strain and rotation waves, as being independent of any real displacements, changes completely our understanding of the discovered appearance of the $E_{\varphi\varphi}$ component in strain measurement related to earthquake events (Gomberg and Agnew 1996; see: Chap. 12).

Additionally we present here some fracture structures due to shear loads (Fig. 1.3), with some micro-rotation processes appearing at both sides of the inner fracture zone. However, in reality the rotations along the slip fragments perpendicular to the main slip plane will be opposite; hence, the resulting rotations would be in such a case almost compensated (Fig. 1.4).

We end this chapter with some experimental evidences related to the rotation motions. We quote the data presented at the two international meetings in California,

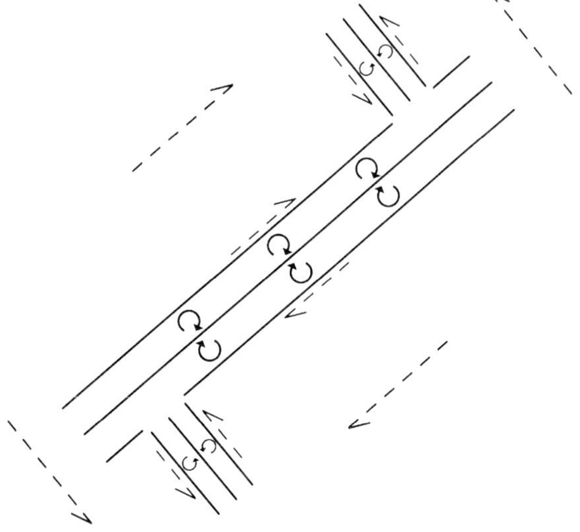

Fig. 1.3 Sketch of slip elements and opposite-sense rotations

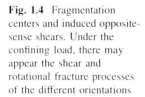

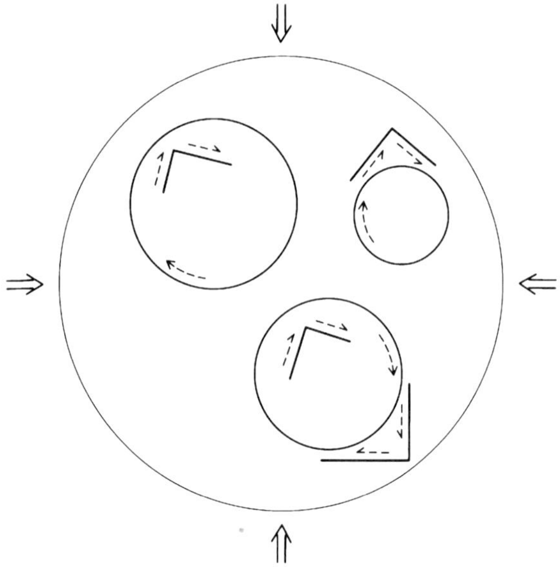

Fig. 1.4 Fragmentation centers and induced opposite-sense shears. Under the confining load, there may appear the shear and rotational fracture processes of the different orientations

USA, 2009 (see: Lee et al. 2009) and Praha, Czech Republic (see: Teisseyre 2012) and additionally we present some experimental observations in Silesia (Poland).

As we have already mentioned, since the beginning of the twenty century there where many attempts to record rotation motions; however, we may note that perhaps the first rotational seismogram has been obtained in an indirect way: the azimuth array of horizontal seismographs, installed in one of the coal mines in Upper Silesia, Poland, to record the very nearby tremors, permitted to deduce the rotation component of motions (see Teisseyre 1974).

Nevertheless, the first consistent records of rotational motions (spin only) are those obtained in the Cashmere Cave, New Zealand and the fundamental geodetic station Wetzell, Germany, at the end of the last and beginning of this century, respectively (e.g., Igel et al. 2005, 2007; Cochard et al. 2006; Schreiber et al. 2006). Here, ring-laser interferometers based on the Sagnac principle were used. These systems, having sensitivity up to 10^{-14} rad/s, were able to record the rotational motions caused by many distant earthquakes. The fiber-optic interferometers, also based on the Sagnac principle, were used by Takeo (2006) for seismic observations in a near field.

T. Moriya (e.g., Moriya and Teisseyre 2006; Teisseyre 2007) has constructed the first rotational seismometer system consisting of a pair of antiparallel seismometers; such a system with a common suspension axis of two antiparallel pendulums was improved in later constructions (Wiszniowski 2006; Wiszniowski et al. 2008). Using the rotation wave records of different events in the very near field (Teisseyre et al. 2003), it was possible to find that for some events, for example the shallow volcanic and the explosion-type ones, we obtain records differing from the common features by the extremely small rotation (spin) portion (Figs. 1.5, 1.6, 1.7).

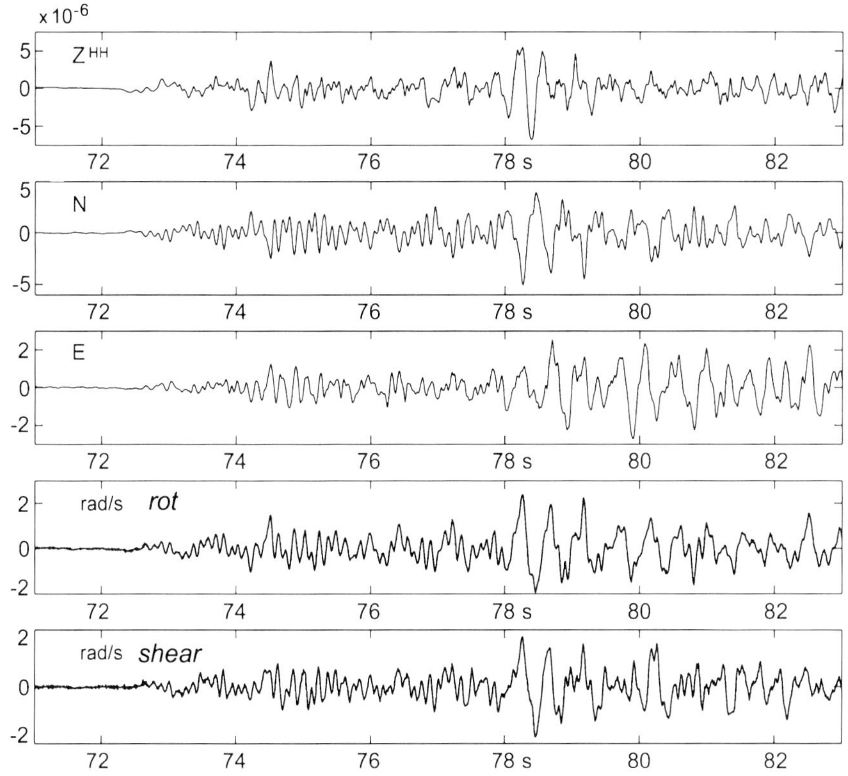

Fig. 1.5 The earthquake in Kaliningrad region, 2004-09-21,11:05:01.6 UT. Velocity seismograms recorded at Ojców Observatory, Poland, 100 Hz sampling. The Pn-wave arrival, time scale adjusted to the time in focus. Z^{HH}—vertical component of the broadband seismometer; other plots—from the rotational seismic station

The P-waves arrival at the Ojców Observatory, time scale adjusted to the time in the focus. Velocity seismograms, 100 Hz sampling. Z^{HH}—vertical component of the broadband seismometer; other plots—from the rotational seismic station (Figs. 1.8, 1.9).

To compare the rotation (spin) and shear (twist) motions, we may use a system of rotational seismometers (Moriya and Marumo 1998; Teisseyre et al. 2003; Teisseyre 2007; Wiszniowski and Teisseyre 2008). Reducing our interest to motions around the vertical axis at the ground surface, we may rely on the data from a couple of perpendicularly oriented rotational seismometers. From each apparatus, which comprises two oppositely oriented horizontal pendulums and two detecting coils, two signals are obtained. Their difference divided by distance between the detecting coils is taken as the differential signal. In this mode, we get two differential signals which both show divergence in horizontal displacement field: first—in one direction, the second—in the perpendicular. From these two

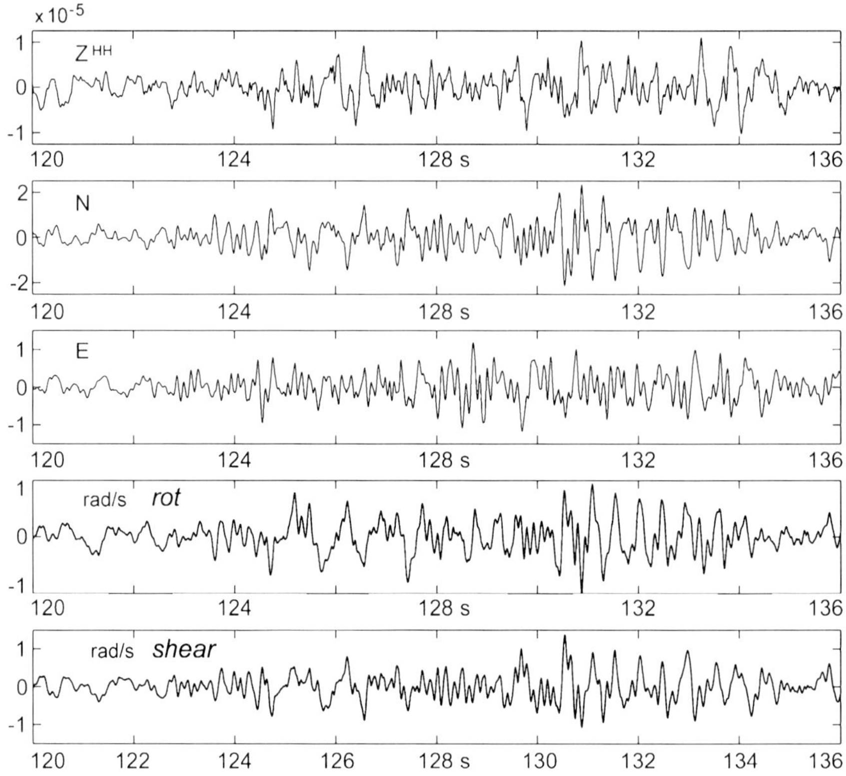

Fig. 1.6 The same event and recordings as in Fig. 1.5. The Sn-wave arrival

signals we can find the spin motion (as the mean value of them) and the shear motion (as half of their difference).

The presented examples of such an analysis, showing rotational components of the seismic fields, were provided by Dr. K. P. Teisseyre.

We should be aware, however, that the shear amplitudes obtained in this way are exact only for the orientation of seismometers coinciding with the main strain axes. Otherwise, the shear data contain an uncertain amplitude scale. Such a system still permits us to compare the recorded spin and shear wavelets as far as their shape in different frequency ranges is concerned. The use of an additional rotational seismometer would allow us to solve the amplitudes problem, if this additional apparatus were positioned in the same plane and at the 45° angle from the axis of one of seismometers of the above-mentioned couple. Thus, the axial motions can be detected with the strainmeters; for small amplitudes we need to use the laser devices.

Having analysed numerous records we observed that both motions (spin and shear) show oscillations of the same of order magnitude (Moriya and Teisseyre 2006; Teisseyre and Suchcicki 2006). Yet we have to remind that investigating the

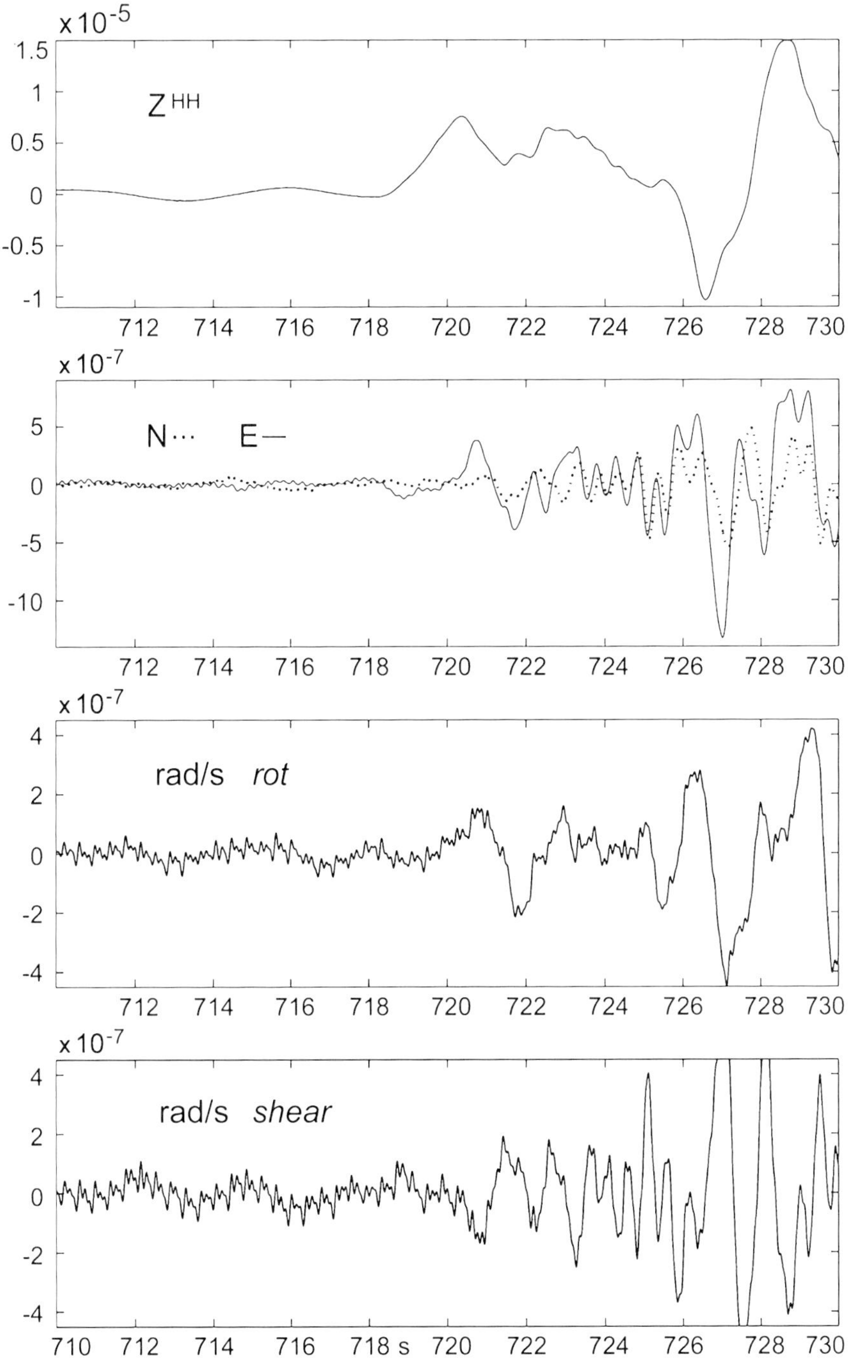

Fig. 1.7 The beginning of Sumatra–Andaman Earthquake, 2004-12-26, 00:58:53.45

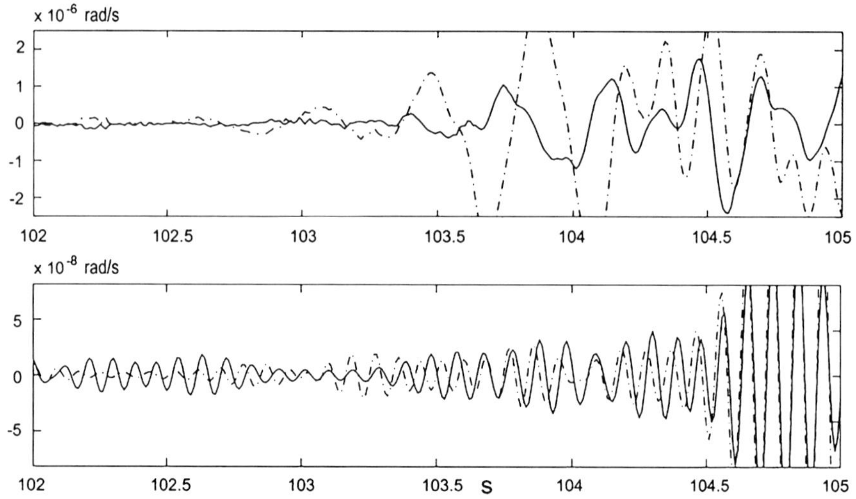

Fig. 1.8 Seismic records of rotations (spin) (*broken lines*) and shear motions (*solid lines*) at the Ioannina Observatory, Greece, 2003-08-14, 05:14:55 UT. The presented time interval, 102–105 s in conventional scale, starts at the moment of P-waves arrival. *Upper panel* spin and shear motions as recorded with two rotational seismometers of perpendicular orientation (the own period of pendulums is 1 s). *Lower panel* the same fields in the frequency interval of 10–12.5 Hz, where the beginning of the event was barely visible

shear field in this way gave only the angular oscillations around the off-diagonal strain axes, while their amplitude was not well defined, depending on the orientation of the measuring system. Only when measuring the shear variations with a system of strainmeters, we can achieve more reliable and independent data. Moreover, the strainmeter system can also measure the axial deformations. It is to be noted that when recording simultaneously the shear and spin, we can derive some new information on the earthquake source processes (see Teisseyre 2009). As we have already noticed the existing technical facilities enable us to record the spin motion (Cochard et al. 2006). Also, the micro-fractures under confining load release the rotation motions (spin) that distinctly exceed the displacement part.

At the end of this Introduction we should underline that one of the important aims of this presentation is to promote organization of the word-wide strain-meter network including both the shear, rotation and axial strain sensors; the existing local networks measuring axial strains in some seismic active regions cannot be helpful to provide a reliable background for the seismic prediction efforts. Of course, we should remember that in the seismic prediction programs very important are records of the non-mechanical fields, like the electric and magnetic ones. However, a main factor causing an increase of seismic risk is related to propagation of the extremely long strain waves; the worldwide observation at network related to the very long axial, shear and rotation strains might become a great challenge for the prediction programs.

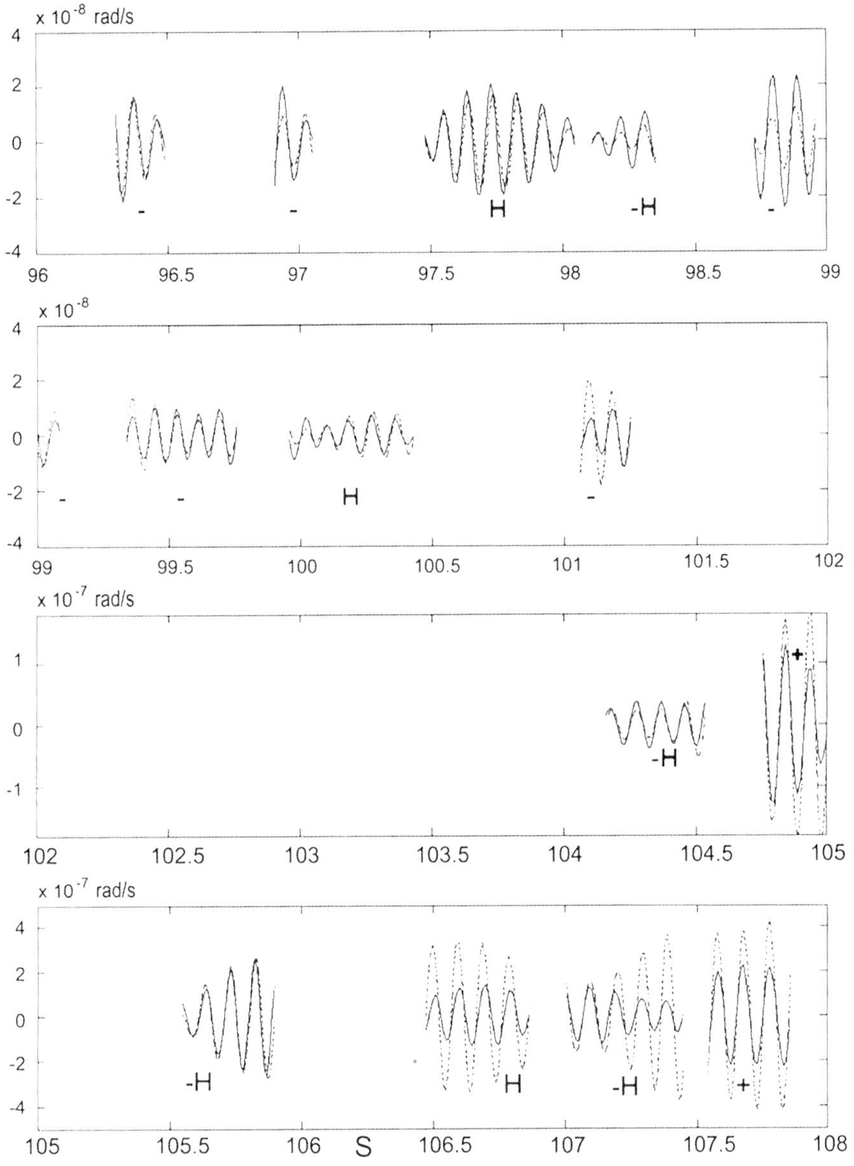

Fig. 1.9 The fragments showing coincidences between the original spin records (*broken lines*) and the original and transformed shear (*solid lines*). The same event as in Fig. 1.4, frequency range 7.5–10 Hz. Four time intervals, as follows: before the event: 96–99 s and 99–102 s, and just after beginning of its recording: 102–105 s and 105–108 s. The fragments with conformance of spin with original shear records are marked with symbol +; when the conformance is with shear wave reversed in sign - with symbol -, while those after the Hilbert transformation – with H (shear phase-shifted by $\pi/2$) and $-$H (shear phase-shifted by $-\pi/2$)

We would like to underline once again that our considerations on the basic elements of the asymmetric continuum theory will lead to a new definition of solid and fluid continua. The basic relations for both continua become basically similar, while their differences depend on the opposite appearance of the molecular motions; such motions become related either to the strains or to the transport fields. We will also get the fracture relations and include the considerations on the non-mechanical effects, e.g., electric or magnetic.

The Asymmetric Strain Theory may become very helpful to a better understanding of some extreme processes, like, e.g., vorticity and turbulence.

Problems of vortex processes observed in geological structures were the subject of book "Vortex-related events of geological processes" where the origin and physics of vertical motions were considered ("Vortex-related events of geological processes" 2004, ed. A.V. Vikulin). The vortex problem is also the subject of more practical study when rotation deformation related to earthquakes creates serious building destruction (Anosov et al. 2004).

Rotation motions play an important role in global tectonogenesis and allow to explain many phenomena observed in nature. The displacement of paleomagnetic and paleoclimatic poles can be explained by relative rotations of the Earth zones around the core. In rotation processes of the Earth main part is played by formation of vertical currents of plastic materials of mantle as a result of convection and eddy currents of deep fluids. Local rotational movements and rotations of continents and lithospheric blocks are explained by such vortex streams. Another phenomena—spherical magnetic anomalies—can be also explained by existence of rotation motions in the Earth crust (Pavlenkova 2007).

References

Anosov GI, Konstantinova TG, Deleman IF (2004) Some evidences relating to the rotation deformations due to the earthquakes as an object of the seismic zonation methods and of the building enhancing problems. In: Vikulin AV (ed) Vortex-related events of the geological processes, Kamchatsky University, Petropavlovsk Kamchatsky

Bykov VG (2006) Solitary waves in crustal faults and their application to earthquakes. In: Teisseyre R, Takeo M, Majewski E (eds) Earthquake source asymmetry, structural media and rotation effects. Springer, Berlin, pp 241–254

Bykov VG (2008) Stick-slip and strain waves in the physics of earthquake rupture: experiments and models. Acta Geophys 56:270–285

Cochard A, Igel H, Schuberth B, Suryanto W, Velikoseltsev A, Schreiber U, Wassermann J, Scherbaum F, Vollmer D (2006) Rotational motions in seismology: theory, observation, simulation. In: Teisseyre R, Takeo M, Majewski E (eds) Earthquake source asymmetry, structural media and rotation effects. Springer, Berlin, pp 391–411

Cosserat E, Cosserat F (1909) Theorie des Corps Déformables. Hermann, Paris

Eringen AC (1999) Microcontinuum field theories i, foundations and solids. Springer, Berlin, p 325

Eringen AC (2001) Microcontinuum field theories II, fluent media. Springer, Berlin, p 340

Eringen AC, Suhubi ES (1964) Non-linear theory of simple microelastic solids-1. Int J Eng Sci 2:189–203

Ferrari G (2006) Note on the historical rotation seismograph. In: Teisseyre R, Takeo M, Majewski E (eds) Earthquake source asymmetry, structural media and rotation effects. Springer, Berlin, pp 367–376
Gomberg J, Agnew DC (1996) The accuracy of seismic estimates of dynamic strains from Pinyon Flat Observatory, California, strainmeter and seismograph data. Bull Seismol Soc Am 86:212–220
Gutenberg B (1926) Physics of the Earth's Interior. Academic Press, New York, pp 111–113
Igel H, Schreiber U, Flaws A, Schuberth B, Velikoseltsev A, Cochard A (2005) Rotational motions induced by the M 8:1 Tokachi-oki earthquake, September 25, 2003. Geophys Res Lett 32:L08309. doi:10.1029/2004GL022336
Igel H, Cochard A, Wassermann J, Flaws A, Schreiber U, Velikoseltsev A, Pham Dinh N (2007) Broadband observations of earthquake induced rotational ground motions. Geoph J Int 168:192–196
Jones WL (1973) Asymmetric wave-stress tensors and wave spin. J Fluid Mech 58:737–747
Kleinert H (1989) Gauge fields in condensated matter, vol II., Stresses and defectsWorld Scientific, Singapore
Kozák JT (2006) Development of earthquake rotational effect study. In: Teisseyre R, Takeo M, Majewski E (eds) Earthquake source asymmetry, structural media and rotation effects. Springer, Berlin, pp 3–10
Kröner E (1981) Continuum theory of defect. In: Balian R, Kleman M, Poirer JP (eds) Les Houches, session xxxv, 1980, physics of defects. North Holland, Amsterdam
Lee WHK, Çelebi M, Todorowska MI, Igel H (2009) Introduction to the special issue on rotational seismology and engineering applications. Bull Seismol Soc Am 99(2B):945–957
Mikhailov DN, Nikolaevskii VN (2000) Tectonic waves of the rotational type generating seismic signals. Izvestiya Phys Sol Earth 36(11):895–902
Mindlin RD (1965) On the equations of elastic materials with microstructure. Int J Solids Struct 1:73
Moriya T, Marumo T (1998) Design for rotation seismometers and their calibration, Hokkaido University. Geophys Bull 61:99–106
Moriya T, Teisseyre R (2006) Design of rotation seismometer and non-linear behaviour of rotation components of earthquakes. In: Teisseyre R, Takeo M, Majewski E (eds) Earthquake source asymmetry, structural media and rotation effects. Springer, Berlin, pp 439–450
Nowacki W (1986) Theory of Asymmetric Elasticity. PWN and Pergamon Press, Warszawa, p 383
Pavlenkova NI 2007. Rotational movements of large elements of the Earth and global geodynamics. In: Rotational processes in geology and physics. Moscow URSS, pp103–115
Schreiber KU, Stedman GE, Igel H, Flaws A (2006) Ring laser gyroscopes as rotation sensors for seismic wave studies. In: Teisseyre R, Takeo M, Majewski E (eds) Earthquake source asymmetry, structural media and rotation effects. Springer, Berlin, pp 377–389
Shimbo M (1975) A geometrical formulation of asymmetric features in plasticity, Hokkaido University. Bull Fac Eng 77:155–159
Shimbo M (1995) Non-Riemannian geometrical approach to deformation and friction. In: Teisseyre R (ed) Theory of earthquake premonitory and fracture processes. PWN (Polish Scientific Publishers), Warszawa, pp 520–528
Takeo M (2006) Ground rotational motions recorded in near-source region of earthquakes. In: Teisseyre R, Takeo M, Majewski E (eds) Earthquake source asymmetry, structural media and rotation effects. Springer, Berlin, pp 157–167
Teisseyre R (1973) Earthquake processes in a micromorphic continuum. Pure Appl Geophys 102:15–28
Teisseyre R (1974) Symmetric micromorphic continuum: wave propagation, point source solutions and some applications to earthquake processes. In: Thoft-Christensen P (ed) Continuu mechanics aspects of geodynamics and rock fracture mechanics. D. Reidel publishing company, Dordrecht, pp 201–244

Teisseyre R (1995) Deformation and geometry. In: Teisseyre R (ed) Theory of Earthquake Premonitory and Fracture Processes. PWN (Polish Scientific Publishers), Warszawa, pp 504–511
Teisseyre KP (2007) Analysis of a group of seismic events using rotational omponents. Acta Geophys 55:535–553
Teisseyre R (2008a) Introduction to asymmetric continuum: dislocations in solids and extreme phenomena in fluids. Acta Geophys 56:259–269
Teisseyre R (2008b) Asymmetric Continuum: Standard Theory. In Teisseyre R, Nagahama H, and Majewski E (eds) Physics of asymmetric continua : extreme and fracture processes,Springer, p95 – 109
Teisseyre R (2009) Tutorial on new development in physics of rotation motions. Bull Seismol Soc Am 99(2B):1028–1039
Teisseyre R (2011) Why rotation seismology: confrontation between classic and asymmetric theories. Bull Seismol Soc Am 101(4):1683–1691
Teisseyre R (2012) Rotation and strain seismology. J Seismology 16(4):683–694. doi:10.1007/s10950-012-9287-6
Teisseyre R, Boratyński W (2003) Continua with self-rotation nuclei: evolution of asymmetric fields. Mech Res Commun 30:235–240
Teisseyre R, Boratyński W (2006) Deviations from symmetry and elasticity: asymmetric continuum mechanics. In: Teisseyre R, Takeo M, Majewski E (eds) Earthquake source asymmetry, structural media and rotation effects. Springer, p31 – 42
Teisseyre R, Górski M (2007) Physics of basic motions in asymmetric continuum. Acta Geophys 55:119–132
Teisseyre R, Górski M (2009a) Fundamental deformations in asymmetric continuum: Motions and fracturing. Bull Seismol Soc Am 99(2B):1028–1039
Teisseyre R, Górski M (2009b) Transport in fracture processes: fragmentation and slip. Acta Geophys 57(3):583–599
Teisseyre R, Górski M, Teisseyre KP (2006) Fracture-Band Geometry and Rotation Energy Release. In: Teisseyre R, Takeo M, Majewski E (eds) Earthquake source asymmetry, structural media and rotation effects, Springer, Berlin, p169 – 184
Teisseyre R, Suchcicki J, Teisseyre KP, Wiszniowski J, Palangio P (2003) Seismic rotation waves: basic elements of the theory and recordings. Ann of Geophys 46:671–685
Teisseyre KP, Suchcicki J (2006) Rotation motions: recording and analysis. In: Teisseyre R, Takeo M, Majewski E (eds) Earthquake source asymmetry, structural media and rotation effects. Springer, Berlin, pp 185–198
Vikulin AV (ed) (2004) Vortex- related events of geological processes, Kamchatsky University, Petropavlovsk Kamchatsky
Wiszniowski J (2006) Rotation and twist motion recording—couple pendulum and rigid seismometers system. In: Teisseyre R, Takeo M, Majewski E (eds) Earthquake source asymmetry, structural media and rotation effects. Springer, Berlin, pp 451–468
Wiszniowski J, Teisseyre R (2008) Field invariant representation: dirac tensors. In: Teisseyre R, Nagahama H, Majewski E (eds) Physics of asymmetric continuum: extreme and fracture processes—earthquake rotation and soliton waves. Springer, Berlin, pp 85–94
Wiszniowski J, Hościłowicz M, Skrzyński A, Suchcicki J, Teisseyre KP (2008) Differential calibration of seismometers for measurement of rotation and strain. Acta Geophys 56:391–407

Chapter 2
Asymmetric Continuum: Basic Motions and Equations

2.1 Introduction

For axial, shear and rotational strains, we present the basic motion equations, which follow directly from the derivatives of the classic Newton formula; these motion equations can be quite independent; their reference displacements can be quite different. However, for the waves emitted from a common source (e.g., an earthquake focus) some of these motions might be correlated or phase shifted; these interaction fields explain the wave propagation due the rotation strain and shear strain mutual release-rebound processes. For a solid continuum we introduce the molecular transport motion as nuclei of a possible real transport inside a fracture domain. We consider also the important experimental data related to the appearances of the $E_{\varphi\varphi}$ components (strain measurement by Gomberg and Agnew 1996). Our theory explains these important discoveries. A similar system of the independent motions is introduced for fluids; besides the transport motion we consider the molecular shear strains and also the rotation molecular strains, both as the time derivatives of these strains.

2.2 Solids: Strain Fields and Molecular Transport

First we recall that the seismometer recordings do not prove the existence of displacements (cf., Eqs. 1.1a, b), just the existence of displacement derivatives; the seismometers record the integrated effect of the displacements derivatives. The strain waves could have even a length of some hundreds kilometers for a very big wave period. We assume that in our Asymmetric Continuuam the real displacements, even very small, do not exist; the real displacements appear only in the fracture processes as an integrated recording effect (cf., Chap. 1).

On the other hand, recent seismological observations of rotation waves by the very precise instruments, based on the Sagnac effect, were able to reveal and measure a very small spin, up to 10^{-9} rad/s. These recorded rotation motions are

R. Teisseyre and M. Teisseyre-Jeleńska, *Asymmetric Continuum*,
GeoPlanet: Earth and Planetary Sciences, DOI: 10.1007/978-3-642-31860-3_2,

not the simple point rotations, but relate to the rotation strains. These experimental observations give grounds for a new, relatively simple theoretical approach in which the strain moments used in a classic theory are replaced by the antisymmetric part of strains. In the Asymmetric Continuum Theory, we use displacements only as the reference motions.

Thus, in solids we can relay on the axial, shear and rotation motions, while the displacements and point rotations are neglected. However, we can add the molecular transport, as possible nuclei for the fracture processes. In this way we arrive at a kind of similarity with the motions in fluids (the transport velocities and the molecular strains). Solids include the strain motions and molecular displacements, while in fluids there will appear the molecular strains and transport motions.

We present the basic strains (axial, shear and rotational) as being related to the independent reference displacements (cf., Teisseyre 2009, 2011); thus all these strains, $E_{ik} = E_{(ik)} + E_{[ik]}$ (or deformations $D_{ik} = E_{ik} = \partial u_k / \partial x_i$) could be released quite independently in a seismic source, or may be mutually related by phase shifted reference displacements, e.g., $\bar{u}_s, \hat{u}_s, \breve{u}_s$:

$$\bar{E}(\bar{u}) = \frac{1}{3}\sum_s E_{ss} = \frac{1}{3}\sum_s \frac{\partial \bar{u}_s}{\partial x_s} \quad - \text{ axial strain} \tag{2.1a}$$

$$\hat{E}_{(ik)}(\hat{u}) = E_{(ik)} - \delta_{ik}\bar{E} = \frac{1}{2}\left(\frac{\partial \hat{u}_k}{\partial x_i} + \frac{\partial \hat{u}_i}{\partial x_k}\right) - \delta_{ik}\frac{1}{3}\sum_{s=1}^{3}\frac{\partial \bar{u}_s}{\partial x_s} \quad - \text{ shear strains} \tag{2.1b}$$

$$\breve{E}_{[ik]}\left(\breve{u}\right) = \frac{1}{2}\left(\frac{\partial \breve{u}_k}{\partial x_i} - \frac{\partial \breve{u}_i}{\partial x_k}\right) \quad - \text{ rotation strain} \tag{2.1c}$$

where the reference displacements may be mutually related, with possible phase shifts, due to the joint source processes; e.g., as follows:

$$\bar{u}_s = \xi^0 u_s^{\%}, \hat{u}_s = e^0 u_s^{\%}, \quad \breve{u}_s = \chi^0 u_s^{\%}; \quad \{\xi^0, e^0, \chi^0\} = \{0, \pm 1, \pm i\} \tag{2.2a}$$

In some situations, the reference displacement may be very useful, e.g., when considering the reflection and refraction rules in the processes related to wave propagation.

We shall note that the presented approach with the independent strain relations (2.1a), or even using the reference displacements, differs essentially from the classic approach in which the solutions might be obtained with the help of a unique displacement field, u (which moreover might be replaced by the potentials, $u = u^P + u^S; \quad u^P = \text{grad } \varphi, \quad u^S = \text{rot } \psi$). Of course, we should remember that a strain rotation, $\breve{E}_{[ik]}$, has a quite different meaning than a simple rotation motion.

The compatibility conditions assure, in a mathematical sense, that the deformations and strains could be expressed by some displacement derivatives:

$$
\begin{aligned}
I_{ij} &= \varepsilon_{ikm}\varepsilon_{jtn}\frac{\partial^2 D_{mn}}{\partial x_k \partial x_t} = 0 \to D_{mn} = \frac{\partial u_n}{\partial x_m} \\
I_{(ij)} &= \varepsilon_{ikm}\varepsilon_{jtn}\frac{\partial^2 E_{(mn)}}{\partial x_k \partial x_t} = 0 \to E_{(mn)} = \frac{1}{2}\left(\frac{\partial u_n}{\partial x_m} + \frac{\partial u_m}{\partial x_n}\right) \\
I_{[ij]} &= \varepsilon_{ikm}\varepsilon_{jtn}\frac{\partial^2 E_{(mn)}}{\partial x_k \partial x_t} = 0 \to E_{[mn]} = \frac{1}{2}\left(\frac{\partial u_n}{\partial x_m} + \frac{\partial u_m}{\partial x_n}\right)
\end{aligned}
\tag{2.2b}
$$

The independent motion equations (cf., Teiseyre and Górski 2009; Teisseyre 2009, 2011) should follow directly from the derivatives of the classic Newton formula:

$$
\begin{aligned}
&\mu\sum_s \frac{\partial^2 u_i}{\partial x_s \partial x_s} - \rho\frac{\partial^2 u_i}{\partial t^2} + (\lambda+\mu)\sum_s \frac{\partial^2 u_s}{\partial x_i \partial x_s} = 0 \text{ and for } D_{ni} = \frac{\partial u_i}{\partial x_n}: \\
&\mu\sum_s \frac{\partial^2 D_{ni}}{\partial x_s \partial x_s} - \rho\frac{\partial^2}{\partial t^2}D_{ni} + (\lambda+\mu)\frac{\partial^2}{\partial x_n \partial x_i}\sum_s D_{ss} = 0
\end{aligned}
\tag{2.3}
$$

We obtain (the external forces omitted):

$$
(\lambda+2\mu)\sum_s \frac{\partial^2 \bar{E}}{\partial x_s \partial x_s} - \rho\frac{\partial^2 \bar{E}}{\partial t^2} = 0; \bar{E} = \frac{1}{3}\sum_s \frac{\partial \bar{u}_s}{\partial x_s} = \frac{1}{3}\sum_s E_{ss} \tag{2.4a}
$$

$$
\begin{aligned}
&\mu\sum_s \frac{\partial^2 \hat{E}_{ni}}{\partial x_s \partial x_s} - \rho\frac{\partial^2}{\partial t^2}\hat{E}_{ni} + (\lambda+\mu)\left(\frac{3\partial^2}{\partial x_n \partial x_i}\bar{E} - \delta_{ni}\sum_s \frac{\partial^2}{\partial x_s \partial x_s}\bar{E}\right) = 0 \text{ or} \\
&\mu\sum_s \frac{\partial^2 \hat{E}_{ni}}{\partial x_s \partial x_s} - \rho\frac{\partial^2}{\partial t^2}\hat{E}_{ni} + (\lambda+\mu)\left(\frac{\partial^2}{\partial x_n \partial x_i}\sum_k E_{kk} - \frac{\delta_{ni}}{3}\sum_{s,k}\frac{\partial^2 E_{kk}}{\partial x_s \partial x_s}\right) = 0
\end{aligned}
\tag{2.4b}
$$

$$
\mu\sum_s \frac{\partial^2}{\partial x_s \partial x_s}\breve{E}_{[ni]} - \rho\frac{\partial^2}{\partial t^2}\breve{E}_{[ni]} = 0 \tag{2.4c}
$$

$$
\begin{aligned}
&E_{kl} = E_{(kl)} + E_{[kl]} = \delta_{kl}\bar{E} + \hat{E}_{kl} + \breve{E}_{[kl]}; \ \bar{E} = \frac{1}{3}\sum_{s=1}^{3} E_{(ss)}, \\
&\hat{E}_{(ik)} = E_{(ik)} - \delta_{ik}\frac{1}{3}\sum_{s=1}^{3} E_{(ss)}
\end{aligned}
\tag{2.4d}
$$

Thus, all these strains (axial, deviatoric, and rotational—related to a load angular moment) could be released, in a common process, with these phase shifts; in that case, these strains will be interrelated through the reference motion, u, as a basic reference field introduced in a mathematical sense. In such a way, we can present the relations for the independent fields with the help of the phase factors, and give their connections to the common reference displacements. In these formulae we have used the different reference displacements and the constitutional relations, joining stresses and strains.

To obtain a relation between the antisymmetric part of stresses, $S_{[kl]}$, and related rotation strains, $E_{[kl]}$, we must introduce the proper constitutive laws, e,g., the Shimbo law (1975, 1995) based on the friction processes and rotation of grains.

Another equivalent constitutive law has been introduced in the Kröner approach to the continuum theory with the self-fields and related internal nuclei (Kröner 1981).

In our approach we join the shear and rotation stresses and strains with common constants μ:

$$S_{(ik)} = 2\mu E_{(ik)} + \lambda\delta_{ik}E_{(ss)} \; and \; \breve{S}_{[ik]} = 2\mu\breve{E}_{[ik]}; \bar{S} = (2\mu + 3\lambda)\bar{E}, \hat{S}_{(ik)} = 2\mu\hat{E}_{(ik)} \tag{2.5a}$$

and

$$\begin{gathered} S_{kl} = S_{(kl)} + S_{[kl]} = \delta_{kl}\bar{S} + \hat{S}_{(kl)} + \breve{S}_{[kl]}; \; \bar{S} = \frac{1}{3}\sum_{s=1}^{3} S_{(ss)}, \; \hat{S}_{(ik)} = S_{(ik)} - \delta_{ik}\frac{1}{3}\sum_{s=1}^{3} S_{(ss)} \\ E_{kl} = E_{(kl)} + E_{[kl]} = \delta_{kl}\bar{E} + \hat{E}_{(kl)} + \breve{E}_{[kl]}; \; \bar{E} = \frac{1}{3}\sum_{s=1}^{3} E_{(ss)}, \; \hat{E}_{(ik)} = E_{(ik)} - \delta_{ik}\frac{1}{3}\sum_{s=1}^{3} E_{(ss)} \\ S_{(ik)} = 2\mu E_{(ik)} + \lambda\delta_{ik}E_{(ss)} \; and \; \breve{S}_{[ik]} = 2\mu\breve{E}_{[ik]}; \; \bar{S} = (2\mu + 3\lambda)\bar{E}, \; \hat{S}_{(ik)} = 2\mu\hat{E}_{(ik)} \end{gathered} \tag{2.5b}$$

The fracture processes in a source can occur due to the release-rebound processes; such interactive processes explain a propagation pattern with the consecutive rotation and shear strains (Teisseyre 1985, 2009, 2011).

When an axial strain is constant, then the release-rebound system may be described by the linear relations between the time and space derivatives; the related equations remind the Maxwell equations. However, to this end we must choose the coordinate system oriented in a special way which presents the deviatoric strains as the off-diagonal tensor:

$$\hat{E}_{(ik)} = E_{(ik)} - \delta_{ik}\frac{1}{3}\sum_{s=1}^{3} E_{(ss)} \rightarrow \hat{E}_{(ik)} = \begin{bmatrix} 0 & \hat{E}_{(12)} & \hat{E}_{(13)} \\ \hat{E}_{(12)} & 0 & \hat{E}_{(23)} \\ \hat{E}_{(13)} & \hat{E}_{(23)} & 0 \end{bmatrix} \tag{2.6}$$

In this system we may define the shear vector as $\hat{E}_i = \{\hat{E}_{(23)}, \hat{E}_{(31)}, \hat{E}_{(12)}\}$, and the rotation vector $\breve{E}_i$ as $\breve{E}_i = \left\{\breve{E}_{[23]}, \breve{E}_{[31]}, \breve{E}_{[12]}\right\}$; note that when using the 4D approach, the vector $\hat{E}_i$can be defined invariantly (cf., Teisseyre 2009; see Chap. 12 on the 4D Maxwell-like invariant relations).

The release-rebound process means that a break of molecular bonds releases rotation field, $\partial\breve{E}/\partial t$, and then in a rebound motion there appears rot$\hat{E}$; reversely, the release of shears, $\partial\hat{E}/\partial t$, leads to rot$\breve{E}$. Finally, we arrive at relations for the release-rebound processes adequately described by the Maxwell-like relations (see: Chap. 12 and Teisseyre 2009, 2011):

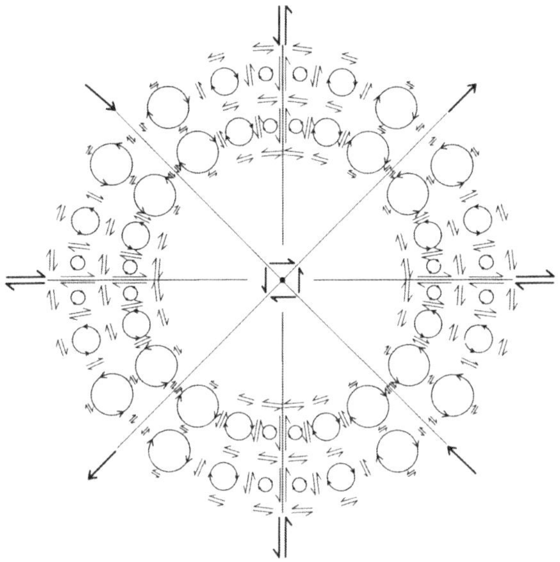

Fig. 2.1 Wave interaction pattern: the shears and rotation strains; their interaction enables propagation of these waves

$$\operatorname{rot} E - \frac{\chi \partial \hat{E}}{c \partial t} = 0, \ \operatorname{rot} \hat{E} + \frac{\chi \partial E}{c \partial t} = 0; \ \frac{c}{\chi} = V^S = \sqrt{\frac{\mu}{\rho}} \tag{2.7a}$$

where c is the light velocity and χ is the material constant.

From these relations we obtain that shear and rotation strain must propagate with the same velocity and that the wave equations which coincide with the previously derived formulae (2.4a), when the axial strains $\sum_s \frac{\partial u_s}{\partial x_s}$ remain to be constant, will be given as:

$$\Delta E - \frac{\chi^2}{c^2} \frac{\partial^2 E}{\partial t^2} = 0 \quad \text{and} \quad \Delta \hat{E} - \frac{\chi^2}{c^2} \frac{\partial^2 \hat{E}}{\partial t^2} = 0 \tag{2.7b}$$

The related wave mosaic (Teisseyre and Gorski 2011) explains the interrelated propagation pattern of the shear and rotation motions as presented in Fig. 2.1 (the rotation strain motions being represented as the double opposite-sense arrows).

The interaction of fields explains the propagation pattern (some critical remarks against rotation motions have been based on the supposition that a separate propagation of rotation motions will be immediately attenuated and therefore could not exist): in our interactive propagation pattern the rotations and shears exist and these motions visualize the wave interactions corresponding to the release-rebound processes at a fracture source.

As mentioned at the beginning of this chapter we should supplement the solid continuum by the molecular transport motion as nuclei of the possible real transport inside a fracture domain, that is outside the considered Asymmetric Continuum. The fracture slip-displacements at a seismic source could develop due to the existence of the assumed molecular displacement velocities, which can relate to the real displacement transport (fracture) forming a fracture zone (formally outside the considered continuum). However, we shall underline that these molecular displacement transports should depend and follow the actual solutions for the strains; in this way the molecular displacements become formed with the help of the real strains and their molecular reference direct displacement transport, or that with the possible phase shifts:

$$d\tilde{\upsilon} \quad \text{or} \quad \left(\xi^0 + e^0 + \chi^0\right) d\tilde{\upsilon} \tag{2.8a}$$

with $\left\{\xi^0, e^0, \chi^0\right\} = \{0, \pm 1, \pm \mathrm{i}\}$ (Eq. 2.2a), as related to the fields $\delta_{kl}\bar{E}, \hat{E}_{(kl)}, E_{[kl]}$ (Eqs. 2.1a–c).

Now, we may define the molecular transport field, w_s, in the following way:

$$w_s = \frac{\partial \upsilon_s}{\partial x_s} \rightarrow \left\{ w_1 = \frac{\partial \upsilon_1}{\partial x_1}, w_2 = \frac{\partial \upsilon_2}{\partial x_2}, w_3 = \frac{\partial \upsilon_3}{\partial x_3} \right\} \tag{2.9a}$$

$$\upsilon = 0, \text{ but } d\upsilon \text{ exists: } d\upsilon = d\frac{\partial u}{\partial t} \tag{2.9b}$$

Here, the derivatives $d\upsilon_s$ mean the derivatives of real transport velocities and the field $w_s = \frac{\partial \upsilon_s}{\partial x_s}$ might become related, after an integration process, to the possible real fracture transport velocities, υ_s, but it appears outside a frame of the continuum.

This molecular transport term should fulfil the Navier–Stokes-like equation:

$$\frac{\mathrm{D}(\rho w_s)}{Dt} = w_s\frac{\partial \rho}{\partial t} + \rho\frac{\partial w_s}{\partial t} + \rho l \sum_k w_k \frac{\partial w_s}{\partial x_k} = \eta \sum_k \frac{\partial^2 w_s}{\partial x_k \partial x_k} \tag{2.10a}$$

where a dynamic micro-viscosity η (the new material constants) and an additional constant l, are introduced.

Thus, we obtain the governing relation for these molecular transports including possible forces:

$$\rho\frac{\partial w_i}{\partial t} + w_i\frac{\partial \rho}{\partial t} + \rho l \sum_k w_k \frac{\partial w_i}{\partial x_k} - \eta \sum_k \frac{\partial^2 w_i}{\partial x_k \partial x_k} = \tilde{F}_i$$

at ρ = const.:

$$\rho\frac{\partial w_i}{\partial t} + \rho l \sum_k w_k \frac{\partial w_i}{\partial x_k} - \eta \sum_k \frac{\partial^2 w_i}{\partial x_k \partial x_k} = \tilde{F}_i \tag{2.10b}$$

Due to this molecular transport and related fracture mechanism we may expect a possible intensity increase of the wave motions in an active seismic region; we will return to these problems in the next chapter.

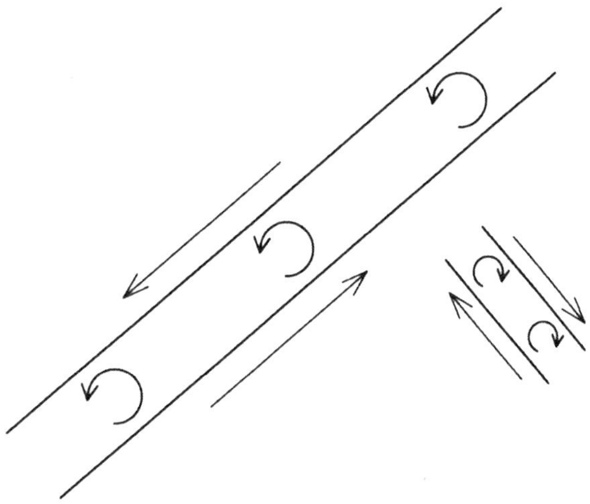

Fig. 2.2 Shear load: sketch of slip elements and the opposite rotations as appear along the main slip and a double couple partner

We may remark that in a similar way to (2.9a), we could define the molecular displacements:

$$w_s^u = \frac{\partial u_s}{\partial x_s} \to \left\{ w_1^u = \frac{\partial u_1}{\partial x_1}, w_2^u = \frac{\partial u_2}{\partial x_2}, w_3^u = \frac{\partial u_3}{\partial x_3} \right\} \text{at} \sum_n w_n^u = 0 \tag{2.11a}$$

and this relation leads us to the equation similar to eqs. (2.4a):

$$\rho \frac{\partial w_n^u}{\partial t} - \mu \sum_k \frac{\partial^2 w_n^u}{\partial x_k \partial x_k} = \ddot{F}_n^u \tag{2.11b}$$

These molecular fields may lead to some fracture process, see, e.g., Fig. 2.2.

Also the molecular transport could lead to an appearance of an interaction chain, the string chain; e.g., such phenomena might be expected near some structure inside the Earth which effectively reflect the seismic waves and also modify the micro-seismic waves. These reflected micro-seismic waves become modified due to the local molecular fields. In this way, the observed reflected micro-seismic parts might bring some earthquake precursory signals as discovered by Sobolev and Lyubushin.

In our final statement we underline that the possible independent fields (displacements, transports, strains and molecular strains) and the different possible float transport motions may be treated as quite independent fields; this is our main postulate. However, it is to be remembered that in some circumstances these fields might be mutually related and shifted in phases; e.g., due to the common fracture processes in a seismic source.

As an example of the powerful abilities of the Asymmetric Theory let us consider the single and double couples. First, let us note the following properties of single and double couples:

- A double couple does not contain any rotational part
- A single couple contains shears and rotation
- Series of the perpendicular couple pairs form rotations with mutually compensated shears.

A double couple field is usually assumed to model a fault slip mechanism (Knopoff and Gilbert 1960); however, when a fault zone has a finite thickness there appears an additional term, the single couple (Knopoff and Chen 2009); the appearing torque imbalance is compensated by a rotation component at the crack tip and therefore the symmetry of Classic Theory is maintained. Nevertheless, the obtained results indicate that the radiation from an advancing crack tip and related strength weakening zone may explain a number of observed rotational phenomena; an additional constitutive law for friction properties should be included.

However, note that in the Asymmetric Continuum Theory a single couple can be defined directly as a sum of the symmetric deviatoric strains and the antisymmetric rotations, $D_{ks} = E^D_{(ks)} + \omega_{ks}$; the related rotation effects remain incorporated into the theory's structure and the theoretical results could be directly related to the observed rotation processes; for the related fracture process we should again include the friction constitutive law.

In classic theory we may consider the solutions for displacements (or displacement potentials) and then estimate a deformation $D_{ks} = \frac{\partial u_s}{\partial x_k}$; otherwise, we can find this solution from the wave equations, and for rotations and deviatoric strains, when assuming a constant confining strain, and putting $F_{nl} = F_{(nl)} + F_{[nl]}$, we will obtain:

$$D_{nl} = E^D_{(nl)} + \omega_{nl},\ \omega_{nl} \equiv E_{[nl]};\ \mu \frac{\partial^2}{\partial x_k \partial x_k} D_{nl} - \rho \frac{\partial^2}{\partial t^2} D_{nl} = F_{nl} \tag{2.12a}$$

The same relation can be obtained in the asymmetric theory, but in that case the fields $E^D_{(nl)}$ and ω_{nl} may be independent or mutually related:

$$D_{nl} = E^D_{(nl)} + \omega_{nl} = \frac{1}{2}\left(\frac{\partial u_l}{\partial x_n} + \frac{\partial u_n}{\partial x_l}\right) - \frac{\delta_{nl}}{3}\frac{\partial u_s}{\partial x_s} + \chi \frac{1}{2}\left(\frac{\partial u_l}{\partial x_n} + \frac{\partial u_n}{\partial x_l}\right);\ \chi = \pm i \tag{2.12b}$$

where this definition of deformation may differ from that given in relation (2.11a).

We may consider the solutions for displacements (or displacement potentials) and then derive the rotation wave and the deviatoric strains; this will be the classic approach. Otherwise, using the asymmetric approach we can solve directly the independent wave equations for the rotation and deviatoric strain fields (with a certain assumption related the axial strain).

These two possible procedures will lead us to quite different results. This is a similar difference between the classic and asymmetric approaches as discussed already in Chap. 1, when considering the problem how to derive the $E_{\varphi\varphi}$ component.

We arrive at some conclusions: confrontation between the Classic Elasticity and Asymmetric Continuum Theory is related to the basic point motions involved; in the former we use exclusively the displacement field, while the asymmetry includes the four point motions: molecular displacements, rotations, deviatoric shears, and axial scalar field. Our consideration indicates an essential difference between the Classic and Asymmetric approaches. Underlining such differences we have pointed out their advantages and deficiencies, but an important fact is that these two theoretical approaches refer, in fact, to the very different technical and measurement systems.

A basic difference between the considered theories follows from the assumption that in the Asymmetric Theory we may admit a simultaneous appearance of a number of fields independently released in a source. This is due to the fact that we admit an independent release of some physical fields, e.g., strains and rotations, in an earthquake source. Formally these released fields might be expressed again by some displacements and, therefore, the resulting rotation and strain fields become related to a number of independent fields. However, the special analysis may permit to separate some observed wave groups originated due to the interaction processes between the rotation and strain fields.

Finally, we shall add that the measurement devices register of course a sum of all possible contributions to a given field, e.g., for displacement we measure a sum of all its contributions from the different source-release processes jointly. Only additional information may discriminate an origin of some component contribution, e.g., for displacement it might be a separately found time period when a wave group of displacement correlates with rotational oscillations (Teisseyre 2007), while for the strain component $E_{\varphi\varphi}$ it is a fact that the displacement-related contribution to that field appears negligible in comparison to the contributions caused by a direct strain effect, as discussed in Chap. 1.

At the end of these considerations let us analyse one special case related to the natural earthquake events in rocks that are under a vertical gradient of pressure; here, it is convenient to use the cylindrical system (r,ϕ,z), with the z-axis oriented in vertical direction. In the considered case we assume the angular bond deformations described by the angular squeeze, $E_{\varphi\varphi}$, in the horizontal plane. Under a significant compression load there may appear some centers with micro-breaks and induced opposite-sense shears (Teisseyre et al. 2006). Such processes can be understood in the following way: the defects become activated under compression load and produce centers with shears of the opposite sense. In effect we observe a rock fragmentation and rotation release followed by the rebound slips. The centers of the opposite sense shears may create dynamic angular deformations leading to the bond breaks and slip propagation followed by the rebound rotations retarded in phase. We can assume that the induced shears and fragmentation depend on the applied load and defect content (cf., Chap. 7). Such induced angular squeeze, $E_{\varphi\varphi}$, can be related to the applied axial stresses due to the defect co-action:

$$\sum S_{ss} = \varepsilon E_{\varphi\varphi} \tag{2.13a}$$

where we insert a new constant, ε, which may relate to the unknown angular squeeze material properties and to a specific angular squeeze structure.

Note a difference between the $E_{\varphi\varphi}$ component and the E_{rr}, E_{zz} ones; the latter components represent typical compression strains, the $E_{\varphi\varphi}$ component represents a circular squeeze which seems to fit to the local fragmentation processes under a high confining load. Such fragmentations in a circular shape are typical in some material crushing under compression. For processes under constant applied load, $E_{rr} + E_{zz} = \text{const}$, we write after eqs. (2.4a) the wave in the cylindrical coordinates, for the field $E_{\varphi\varphi}$:

$$\mu\left(\frac{\partial}{r\partial r}\left(\frac{r\partial}{\partial r}\right)+\frac{\partial^2}{r^2\partial\psi^2}+\frac{\partial^2}{\partial z^2}\right)E_{\varphi\varphi}-\rho\frac{\partial^2}{\partial t^2}E_{\varphi\varphi}=\left(\frac{\mu}{\lambda+2\mu}\right)^{1/2}\frac{\partial F_\varphi}{r\partial\psi} \tag{2.13b}$$

where $\psi = \varphi\left(\frac{\mu}{\lambda+2\mu}\right)^{1/2}$

Considering the angular squeeze deformations related to the independent axial strains, we get a completely new view on the appearance of the $E_{\varphi\varphi}$ component discovered in strain measurement related to earthquake events (Gomberg and Agnew 1996). In these authors' original opinion, related to the Classical Elasticity, the squeeze strain angular components, $E_{\varphi\varphi}$, can be estimated from the solution for the displacement potentials. A pressure gradient (z-direction) may have an angular symmetry of a squeeze appearing in a horizontal plane (parallel to the Earth surface). These authors, using the scalar and vector potentials, Φ and Ψ, considered the asymptotic solution for the cylindrical waves excited by an earthquake source:

$$(\Phi,\Psi)\rightarrow f(r,\varphi,t)=AJ_m(kr)\exp[\mathrm{i}(m\varphi-\varpi t)], \tag{2.14a}$$

where for m = 0 we obtain:

$$f(r,t)\approx A\sqrt{\frac{2}{\pi k_r r}}\exp[\mathrm{i}(k_r r-\varpi t-\pi/4)] \tag{2.14b}$$

The exact solutions would be given by an expansion of the Bessel functions with cylindrical harmonics (cf., Udias 2002).

An asymptotic expression for displacements, and further for the strain angular squeeze $E_{\varphi\varphi}$, becomes as follows (Gomberg and Agnew 1996):

$$E_{\varphi\varphi}\approx\frac{u_r}{r}+i\frac{u_\varphi}{r} \tag{2.14c}$$

The approximated solution decays rapidly with distance; the cylindrical wave front may be estimated as close to zero. These authors wanted to explain the experimentally discovered angular squeeze strains discovered by the systematic squeeze measurements. The obtained theoretical result (2.14c) forced them to assume that the experimentally discovered angular squeeze strains appear only as the local distortions of the strain field, in spite of the fact that for all analysed events the obtained $E_{\varphi\varphi}$values have been estimated as significant.

However, when we return to the relations valid in the Asymmetric Continuum Theory, an interpretation of the above-mentioned results look quite differently. There appears a big difference between the Classical approach of an angular squeeze, derived from the wave equation for displacement potentials, and the Asymmetric Theory considering directly the theoretical solution for the independent squeeze waves. In this case instead of the solution (2.14c) we arrive at the Bessel wave solution for this angular term:

$$E_{\varphi\varphi} = AJ_m(kr)\exp[\mathrm{i}(k_z z + m\psi - \varpi t)]; \quad \psi = \varphi\left(\frac{\mu}{\lambda + 2\mu}\right)^{1/2} \tag{2.15a}$$

In first approximation the asymptotic expression for $E_{\varphi\varphi}(r, \psi, z, t)$ can be given by an expansion of the Bessel functions:

$$\begin{aligned} E_{\varphi\varphi} \propto & \sqrt{\frac{2}{\pi k_r r}}\exp[\mathrm{i}(k_r r + k_z z + m\psi - \varpi t - \pi/4)] \approx \\ & \sqrt{\frac{2}{\pi k_r r}}\exp\left[\mathrm{i}\left(k_r r + k_z z + \frac{m\varphi}{\sqrt{3}} - \varpi t - \pi/4\right)\right] \end{aligned} \tag{2.15b}$$

In this solution, the angular squeeze decays with r quite differently than in the asymptotic expression (2.14c) derived from the Classic Theory: this difference follows from the fact that we rely on an independent wave equation for all strains, including the squeeze ones. The related wave solution represents the propagation of the circular squeeze deformation $E_{\varphi\varphi}(r, z)$; such an independent physical field can propagate due the asymmetric continuum structure exciting much greater effects than those previously estimated by Gomberg and Agnew (1996).

Concluding, we can treat the observed deformations, $E_{\varphi\varphi}$, as real experimental facts explained by the Asymmetric Theory; the obtained result leads to an agreement between observations and theory. We repeat that the angular squeeze strains considered by Gomberg and Agnew (1996) were recorded as significant for all the events they analysed.

Further, let us assume a micro-fracture at r = 0 under the condition $E_{rr} + E_{\varphi\varphi} = \text{constant}\ (E_{rr} = -E_{\varphi\varphi})$, then the wave oscillations of these components become given by the equations (cf. Eq. 2.4a):

$$\mu\Delta E_{rr} - \rho\frac{\partial^2}{\partial t^2}E_{rr} = \frac{\partial F_r}{\partial r}, \mu\Delta E_{\varphi\varphi} - \rho\frac{\partial^2}{\partial t^2}E_{\varphi\varphi} = \frac{\partial F_\varphi}{r\partial\varphi} \tag{2.16a}$$

This will lead to the wave solution for a radial component, E_{rr}, and the Bessel solution for squeeze, $E_{\varphi\varphi}$, strains (Fig. 2.3.); in approximation:

$$E_{\varphi\varphi} \propto \sqrt{\frac{2}{\pi k_r r}}\exp[\mathrm{i}(k_r r + m\varphi - \varpi t - \pi/4)] \tag{2.16b}$$

The next figure, Fig. 2.4, presents a combined propagation of these deformations, E_{rr} E_{rr} and $E_{\varphi\varphi}$, with the P-wave velocity: $V^P = \varpi/k_r$.

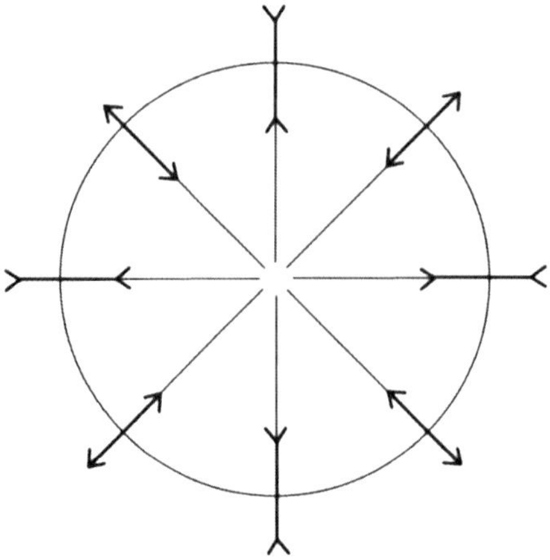

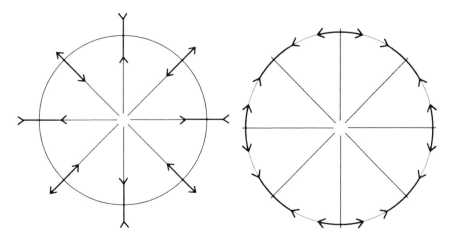

Fig. 2.3 Sketch of the E_{rr} deformations (*left part*) and $E_{\varphi\varphi}$ deformations (*right part*) related to the solutions for m = 2

Fig. 2.4 Propagation related to the interaction of the E_{rr} and $E_{\varphi\varphi}$ strains

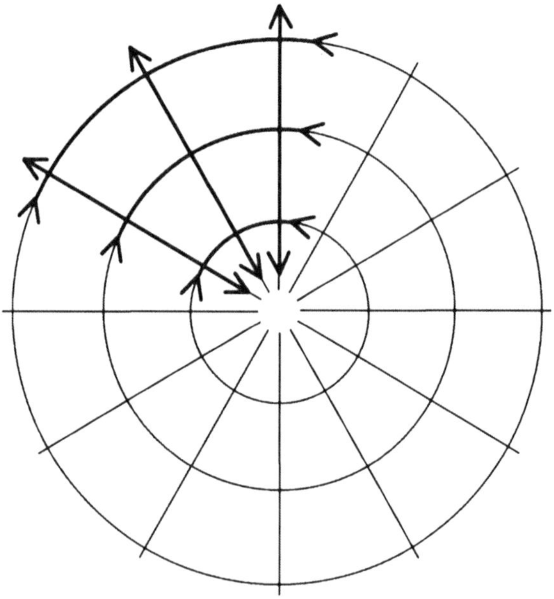

2.3 Fluids: Molecular Strains and Transport

A similar asymmetric approach can be introduced for fluids; we also postulate that the different motion processes in fluids may either remain as the quite independent fields, or be mutually interdependent.

A transport motion in fluids is described by the Navier–Stokes transport equations. To consider the transport processes we recall the Euler equation

$$\frac{\partial \upsilon}{\partial t} + (\upsilon \nabla)\upsilon = -\frac{1}{\rho}\,\mathrm{grad}\,p + g \tag{2.17}$$

The transport processes at a constant density can be presented symbolically by the following transition:

$$\frac{\partial}{\partial t} \rightarrow \frac{\mathrm{d}}{\mathrm{d}t} = \frac{\partial}{\partial t} + \sum_s \upsilon_s \frac{\partial}{\partial x_S} \tag{2.18}$$

leading to the Navier–Stokes equation at a constant density:

$$\rho \frac{\mathrm{d}\upsilon_i}{\mathrm{d}t} = \rho \frac{\partial \upsilon_i}{\partial t} + \rho \sum_s \upsilon_s \frac{\partial \upsilon_i}{\partial x_s} = \eta \sum_s \frac{\partial^3 \upsilon_i}{\partial x_s^2} + F_i \tag{2.19}$$

where F are the body forces, υ is the displacement velocity, η is the dynamic viscosity, and p is the pressure.

Now we note that these transport motions, υ_i, should influence directly the molecular strain fields, $\tilde{E}_{(ki)}$ and $\tilde{E}_{[ki]}$, as a kind of their reference field. The molecular stresses and strains present the symmetric and anti-symmetric molecular deformations (Teisseyre 2009):

$$\tilde{S}_{kl} = \tilde{S}_{(kl)} + \tilde{S}_{[kl]}, \tilde{E}_{kl} = \tilde{E}_{(kl)} + \tilde{E}_{[kl]} \tag{2.20a}$$

Here, we refer to some of our former papers (Teisseyre 2007, 2008, 2009); these fields can be inter-related by means of the following constitutive relations

$$\frac{1}{3}\tilde{S}_{ss} = -p = \frac{1}{3}k\sum_s \tilde{E}_{ss}, \quad \bar{\bar{S}} = \frac{1}{3}k\sum_s \bar{\tilde{E}}_{ss} = k\bar{\bar{E}};$$
$$\hat{\tilde{S}}_{(kl)} = \tilde{\mu}\hat{\tilde{E}}_{(kl)}, \tilde{S}_{[kl]} = \tilde{\mu}\tilde{E}_{[kl]}; \hat{\tilde{S}}_{(kl)} = \tilde{S}_{(kl)} - \frac{1}{3}\delta_{kl}\tilde{S}_{ss}, \hat{\tilde{E}}_{(kl)} = \tilde{E}_{(kl)} - \frac{1}{3}\delta_{kl}\tilde{E}_{ss} \tag{2.20b}$$

where $\hat{\tilde{S}}_{(kl)}$ and $\hat{\tilde{E}}_{(kl)}$ mean the deviatoric parts of the symmetric tensors and $\tilde{S}_{[kl]}$, $\tilde{E}_{[kl]}$ relate to antisymmetric molecular tensors; for both constitutive relations we have assumed the same rigidity constant $\tilde{\mu}$ (this assumption is supported by the similar arrivals of the shear and rotation motions).

The molecular shear strain rate and molecular shear rotation rate (molecular spin) can be related to some reference displacement velocities in the following way:

$$\tilde{E}_{(kl)} = \frac{1}{2}\left(\frac{\partial \bar{\tilde{\upsilon}}_l}{\partial x_k} + \frac{\partial \bar{\tilde{\upsilon}}_k}{\partial x_l}\right), \ \breve{E}_{[kl]} = \frac{1}{2}\left[\frac{\partial \breve{\upsilon}_l}{\partial x_k} - \frac{\partial \breve{\upsilon}_k}{\partial x_l}\right], \ \bar{\bar{E}} = \frac{1}{3}\sum_s \frac{\partial \bar{\bar{\upsilon}}_S}{\partial X_S}$$
$$\hat{\tilde{E}}_{(ik)} = \tilde{E}_{(ik)} - \delta_{ik}\bar{E} = \frac{1}{2}\left(\frac{\partial \hat{\upsilon}_k}{\partial x_i} + \frac{\partial \hat{\upsilon}_i}{\partial x_k}\right) - \delta_{ik}\frac{1}{3}\sum_{s=1}^{3}\frac{\partial \bar{\bar{\upsilon}}_S}{\partial x_S} \tag{2.21}$$

where the relation for antisymmetric molecular strains follows from the fact that any rotation field can be expressed as a rotation of some velocity field; the reference velocities $\bar{\bar{\upsilon}}_k, \hat{\upsilon}_k, \breve{\upsilon}_k$ are introduced as mathematical reference fields only.

The relations for the molecular strain follow automatically from those presented for the real strains. We may assume an independence between the introduced molecular strains (strain rates), or these fields could be mutually related due to some internal processes in fluid, e.g., vortex motion.

It shall be noted, also, that for the equal reference motions, $\bar{\bar{\upsilon}}_l, \breve{\upsilon}_l, \tilde{\upsilon}_l$, and for $k = \tilde{\mu}$, we obtain the relation between the molecular stresses and displacement velocity as a sum of molecular strain rate and spin:

$$\tilde{S}_{kl} = \tilde{S}_{(kl)} + \tilde{S}_{[kl]} = \tilde{\mu}\frac{\partial \tilde{\upsilon}_l}{\partial x_k} = \tilde{\mu}\left(\tilde{E}_{(kl)} + \tilde{E}_{[kl]}\right), \ \tilde{S}_{kl} = \tilde{\mu}\tilde{E}_{kl} \tag{2.22}$$

We may consider the balance equation for spin as follows (Teisseyre 2007, 2008, 2009):

$$\tilde{M}_{lk} = l^2 \varepsilon_{lki}\sum_n \frac{\partial \tilde{S}_{[ni]}}{\partial x_n} = \tilde{\mu} l^2 \varepsilon_{lki}\sum_n \frac{\partial \tilde{E}_{[ni]}}{\partial x_n} \tag{2.23a}$$

$$\frac{\partial \tilde{M}_{pk}}{\partial x_k} = l^2\sum_n \varepsilon_{pki}\frac{\partial^2 \tilde{S}_{[ni]}}{\partial x_k \partial x_n} = l^2\sum_n \varepsilon_{pki}\frac{\partial^2 \tilde{S}_{[ki]}}{\partial x_n \partial x_n} = l^2\rho\varepsilon_{pki}\frac{\partial \tilde{E}_{[ki]}}{\partial t} + \varepsilon_{pki}\tilde{K}_{[ki]} \tag{2.23b}$$

where l is the Cosserat length enabling us to formulate the action of molecular stresses and moments on the fluid elements, and $\tilde{K}_p = \varepsilon_{pki}\tilde{K}_{[ki]}$ means the external moment.

We noticed also an important equivalence of these different summations:

$$\sum_n \varepsilon_{pki}\frac{\partial^2 \tilde{S}_{[ni]}}{\partial x_k \partial x_n} = \sum_n \varepsilon_{pki}\frac{\partial^2 \tilde{S}_{[ki]}}{\partial x_n \partial x_n} \tag{2.24}$$

Due to the further transformations we can write

$$\frac{\partial \tilde{M}_{pk}}{\partial x_k} = \tilde{\mu} l^2 \varepsilon_{pks}\sum_n \frac{\partial^2 \tilde{E}_{[ks]}}{\partial x_n \partial x_n} = \rho l^2 \varepsilon_{pks}\frac{\partial \tilde{E}_{[ks]}}{\partial t} + \varepsilon_{pks}\tilde{K}_{[ks]}, \tag{2.25}$$

We may define now an average rotation over some group of the neighbouring rotations considered in an element related to the characteristic length element; we put:

$$\tilde{\mu}\sum_k \frac{\partial^2 \tilde{E}_{[ns]}}{\partial x_k \partial x_k} = \rho \frac{\partial^2 \tilde{E}_{[ns]}}{\partial t^2} \rightarrow \tilde{\mu}\sum_k \frac{\partial^2 \Omega_p}{\partial x_k \partial x_k} = \rho \frac{\partial^2 \Omega_p}{\partial t^2} \tag{2.26}$$

where η, ρ are constants and $\omega_p = \varepsilon_{pks}\tilde{E}_{[ks]}$ and $\Omega_p = \frac{1}{\Delta\omega}\int \omega_p \mathrm{d}\omega$.

There may be some reasons to average the pure spin (molecular rotation strains) into a molecular rotation related to some Cosserat characteristic molecular length, we will return to these relations in the next chapter, however, for a constant density the considered structure as a whole can be attributed to the field Ω (Eq. 2.26), while rotation motions at each of points, of course, may be described by the field ω.

At the end of this chapter we recall the structure of the Kröner theory (Kröner 1981) which allows to introduce additional fields, e.g., the rotation ones. Comparing our assumptions to that theory we note that the Kröner elastic fields correspond to the symmetric fields and the self-fields to the asymmetric fields, while the total fields correspond to the physical fields, which in the Kröner theory are represented by the elastic fields.

We should also remember that instead of stress moments we use an equivalent system representing antisymmetric stresses $S_{[ik]}$; the related balance law expresses the rotation force moment acting on a body element as the antisymmetric stresses (Teisseyre and Boratyński 2003, 2006).

Stress moment M_i can be formed from the stress moment tensor M_{is}. On the basis of the antisymmetric stresses

$$M_i = \frac{\partial}{\partial x_s} M_{is} = \in_{iks} \frac{\partial^2}{\partial x_n \partial x_s} S_{[kn]} \text{ and } M_{is} = \in_{iks} \frac{\partial}{\partial x_n} S_{[kn]} \tag{2.27a}$$

we arrive at

$$\varepsilon_{iks} \frac{\partial^2}{\partial x_n \partial x_s} S_{[kn]} = 2\mu\varepsilon_{iks} \frac{\partial^2}{\partial x_n \partial x_s} E_{[kn]} = \frac{1}{2}\varepsilon_{iks}\rho \frac{\partial^2}{\partial t^2} E_{[kn]}, \tag{2.27b}$$

where rotation might be represented as: $\omega_i = \frac{1}{2}\varepsilon_{iks}E_{[ks]}$.

We may note that according to Shimbo (1975) we use the same modulus for the symmetric and antisymmetric constitutive law joining the stresses and strains (see: Teisseyre et al. 2006, Chaps. 4–6).

In our approach, instead of stress moments we use an equivalent system representing antisymmetric stresses S[ik]; the related balance law expresses, on the one hand, the rotation of force acting on a body element due to the antisymmetric stresses (stress moment divided by an infinitesimal arm length), and, on the other hand, the balancing term, i.e., the acceleration related to angular momentum (Teisseyre and Boratyński 2003, 2006).

We may note that in the micromorphic continuum (Eringen 1999) the microstrain can contain its antisymmetric part while the gyration tensor can contain its symmetric part; these peculiarities extend, moreover, into the stresses and stress moments and on the inertia spin tensor (in the latter, some additional asymmetric properties may result from the micro-inertia tensor).

References

Eringen AC (1999) Microcontinuum field theories I, foundations and solids. Springer, Berlin, p 325

Gomberg J, Agnew DC (1996) The accuracy of seismic estimates of dynamic strains from Pinyon Flat Observatory, California, strainmeter and seismograph data. Bull Seismol Soc Am 86:212–220

Knopoff L, Chen YT (2009) The single couple equivalent force component of dynamical fractures and intrafault rotations. Bull Seismol Soc Amer 50:117–133

Knopoff L, Gilbert F (1960) First motions from seismic sources. Bull Seismol Soc Amer 50:117–133

Kröner E (1981) Continuum theory of defect. In: Les houches, session

Shimbo M (1975) A geometrical formulation of asymmetric features in plasticity, Hokkaido University. Bull Fac Eng 77:155–159

Shimbo M (1995) Non-Riemannian geometrical approach to deformation and friction. In: Teisseyre R (ed) Theory of earthquake premonitory and fracture processes. PWN (Polish Scientific Publishers), Warszawa, pp 520–528

Teisseyre R (1985) New earthquake rebound theory. Phys Earth Planet Inter 39:1–4

Teisseyre KP (2007) Analysis of a group of seismic events using rotational omponents. Acta Geophys 55:535–553

Teisseyre R (2008) Asymmetric continuum: standard theory. In: Teisseyre R, Nagahama H, Majewski E (eds) Physics of asymmetric continua : extreme and fracture processes. Springer, Berlin, p 95–109

Teisseyre R (2009) Tutorial on new development in physics of rotation motions. Bull Seismol Soc Am 99(2B):1028–1039

Teisseyre R (2011) Why rotation seismology: confrontation between classic and asymmetric theories. Bull Seismol Soc Am 101(4):1683–1691

Teisseyre R, Boratyński W (2003) Continua with self-rotation nuclei: evolution of asymmetric fields. Mech Res Commun 30:235–240

Teisseyre R and Górski M (2009) Fundamental deformations in asymmetric continuum: motions and fracturing, Bull Seismol Soc Am 99, 2B:1132–1136

Teisseyre R, Białecki M, Górski M (2006) Degenerated asymmetric continuum theory. In: Teisseyre R, Takeo M, Majewski E (eds) Earthquake source asymmetry, structural media and rotation effects. Springer, Berlin, p 43–56

Chapter 3
Transport and Float Transport Motions

3.1 Introduction

We consider the equations of motion when we take into consideration more than one independent float transport. As an example we consider the Coriolis forces and Eötvös effect, both related to the Earth rotation and some linear transport in fluids; such problems relate to a double system of equations. In solids, a fracture event can be related simultaneously to the shear and rotation fracture process. The molecular transport introduced into the solid continuum plays an essential role in the fracture processes; for this molecular transport we propose the relations similar to the Navier–Stokes equations. For fluids we consider the molecular stresses; further, we follow the Landau and Lifschitz (1959) approach. Finally we discuss a flow motion in glaciers and the related fracture properties: here we combine some processes occurring in solids and fluids, as in glaciers there appear both the fracture events and the ice-vibrations as a mixture of fracture and flow processes.

3.2 Coriolis Effect

The equations of motion become more complicated when we take into consideration more than one independent float transport motion. We usually have the transport motions with mutually related float transports, υ_k and $\tilde{\upsilon}_k$; as an example we may obtain a joint equation system for the fields, e.g., $\{\tilde{\upsilon}_r, \tilde{\upsilon}_\varphi, \tilde{\upsilon}_z\}$ and $\{\upsilon_1, \upsilon_2, \upsilon_3\}$, which appear due to different reasons; thus, we write for the field υ with a float transport $\tilde{\upsilon}$:

$$\rho \frac{\mathrm{d}\upsilon_s}{\mathrm{d}t} = \rho \frac{\partial \upsilon_s}{\partial t} + \rho \sum_k \upsilon_k \frac{\partial \upsilon_s}{\partial x_k} + \rho \tilde{\upsilon}_r \frac{\partial \upsilon_s}{\partial r} + \rho \tilde{\upsilon}_\varphi \frac{\partial \upsilon_s}{r \partial \varphi} + \rho \tilde{\upsilon}_z \frac{\partial \upsilon_s}{\partial z} = \eta \Delta \upsilon_s + F_s \tag{3.1a}$$

R. Teisseyre and M. Teisseyre-Jeleńska, *Asymmetric Continuum*,
GeoPlanet: Earth and Planetary Sciences, DOI: 10.1007/978-3-642-31860-3_3,

and a second equation for the field $\tilde{\upsilon}$ with a float transport υ:

$$\rho\frac{\mathrm{d}\tilde{\upsilon}_r}{\mathrm{d}t} = \rho\frac{\partial\tilde{\upsilon}_r}{\partial t} + \rho\tilde{\upsilon}_r\frac{\partial\tilde{\upsilon}_r}{\partial r} + \rho\tilde{\upsilon}_\varphi\frac{\partial\tilde{\upsilon}_r}{r\partial\varphi} + \rho\tilde{\upsilon}_z\frac{\partial\tilde{\upsilon}_r}{\partial z} + \rho\sum_s\upsilon_s\frac{\partial\tilde{\upsilon}_r}{\partial x_s} = \eta\Delta\tilde{\upsilon}_r + \tilde{F}_r$$

$$\rho\frac{\mathrm{d}\tilde{\upsilon}_\varphi}{\mathrm{d}t} = \rho\frac{\partial\tilde{\upsilon}_\varphi}{\partial t} + \rho\tilde{\upsilon}_\varphi\frac{\partial\tilde{\upsilon}_\varphi}{\partial r} + \rho\tilde{\upsilon}_\varphi\frac{\partial\tilde{\upsilon}_\varphi}{r\partial\varphi} + \rho\tilde{\upsilon}_z\frac{\partial\tilde{\upsilon}_\varphi}{\partial z} + \rho\sum_s\upsilon_s\frac{\partial\tilde{\upsilon}_\varphi}{\partial x_s} = \eta\Delta\tilde{\upsilon}_\varphi + \tilde{F}_\varphi$$

$$\rho\frac{\mathrm{d}\tilde{\upsilon}_z}{\mathrm{d}t} = \rho\frac{\partial\tilde{\upsilon}_z}{\partial t} + \rho\tilde{\upsilon}_r\frac{\partial\tilde{\upsilon}_z}{\partial r} + \rho\tilde{\upsilon}_\varphi\frac{\partial\tilde{\upsilon}_z}{r\partial\varphi} + \rho\tilde{\upsilon}_z\frac{\partial\tilde{\upsilon}_z}{\partial z} + \rho\sum_s\upsilon_s\frac{\partial\tilde{\upsilon}_z}{\partial x_s} = \eta\Delta\tilde{\upsilon}_z + \tilde{F}_z$$

(3.1b)

We obtain a double system of the related equations which for both variable fields should be solved as one combined system; we will return to this system in the chapter related to the turbulence problems. However, we may note that in most cases one of these double transport fields considered remains constant and then the presented system reduces to one set of equations.

For a float transport expressed in a spherical system $\{r, \theta, \psi\}$ we have:

$$\rho\frac{\partial\upsilon_i}{\partial t} + \rho\sum_s\upsilon_s\frac{\partial\upsilon_i}{\partial x_s} + \rho\breve{\upsilon}_r\frac{\partial\upsilon_i}{\partial r} + \rho\breve{\upsilon}_\theta\frac{\partial\upsilon_i}{r\partial\theta} + \rho\breve{\upsilon}_\psi\frac{\partial\upsilon_i}{r\sin\theta\partial\psi} = \eta\sum_s\frac{\partial^2\upsilon_i}{\partial x_s^2} + F_i$$

(3.2a)

and for the second equation related to it we get:

$$\rho\frac{\partial\breve{\upsilon}_r}{\partial t} + \rho\breve{\upsilon}_r\frac{\partial\breve{\upsilon}_r}{\partial r} + \rho\breve{\upsilon}_\theta\frac{\partial\breve{\upsilon}_r}{r\partial\theta} + \rho\breve{\upsilon}_\psi\frac{\partial\breve{\upsilon}_r}{r\sin\theta\,\partial\psi} + \rho\sum_s\upsilon_s\frac{\partial\breve{\upsilon}_r}{\partial x_s} = \eta\Delta\breve{\upsilon}_r + \breve{F}_r$$

$$\rho\frac{\partial\breve{\upsilon}_\theta}{\partial t} + \rho\breve{\upsilon}_r\frac{\partial\breve{\upsilon}_\theta}{\partial r} + \rho\breve{\upsilon}_\theta\frac{\partial\breve{\upsilon}_\theta}{r\partial\theta} + \rho\breve{\upsilon}_\psi\frac{\partial\breve{\upsilon}_\theta}{r\sin\theta\,\partial\psi} + \rho\sum_s\upsilon_s\frac{\partial\breve{\upsilon}_\theta}{\partial x_s} = \eta\Delta\breve{\upsilon}_\theta + \breve{F}_\theta$$

$$\rho\frac{\partial\breve{\upsilon}_\psi}{\partial t} + \rho\breve{\upsilon}_r\frac{\partial\breve{\upsilon}_\psi}{\partial r} + \rho\breve{\upsilon}_\theta\frac{\partial\breve{\upsilon}_\psi}{r\partial\theta} + \rho\breve{\upsilon}_\psi\frac{\partial\breve{\upsilon}_\psi}{r\sin\theta\,\partial\psi} + \rho\sum_s\upsilon_s\frac{\partial\breve{\upsilon}_\psi}{\partial x_s} = \eta\Delta\breve{\upsilon}_\psi + \breve{F}_\psi$$

(3.2b)

We may consider other relative combinations of motion and float, e.g., an interaction of the two rotation fields or interaction of the rotation and cylindrical motions.

A more general superposition of such float transport systems could be related to an influence of more than two flow motions.

Another problem appears when we consider the combined transport floats related to the different material reference systems; in such a case we should take into account different operators related to the combined objects, e.g., the rotating Earth and a river flow in it. For this example, we may consider the Coriolis forces and Eötvös effect; these complex transport motions related to the different reference systems can be tested for the Coriolis force related to the Earth rotation, ω_k, and another relative motion, υ_i:

$$F_s = -2\rho\varepsilon_{ski}\omega_k \upsilon_i\ , \quad a_s^C = -2\begin{Bmatrix} \upsilon_U \cos\psi - \upsilon_N \sin\psi \\ \upsilon_E \sin\psi \\ -\upsilon_E \cos\psi \end{Bmatrix}, \omega_k = \omega\begin{Bmatrix} 0 \\ \cos\psi \\ \sin\psi \end{Bmatrix},$$
$$\upsilon_i = \begin{Bmatrix} \upsilon_E \\ \upsilon_N \\ \upsilon_U \end{Bmatrix} \tag{3.3a}$$

where $\upsilon_E = \upsilon_{East}$; $\upsilon_N = \upsilon_{North}$; $\upsilon_U = \upsilon_{Up}$.

Otherwise, we may write with an acceleration term, $F_s = \rho a_s^C$, or:

$$\begin{Bmatrix} F_E \\ F_N \\ F_U \end{Bmatrix} = \rho \begin{Bmatrix} a_E^C \\ a_N^C \\ a_U^C \end{Bmatrix} = -2\rho \begin{Bmatrix} \upsilon_U \cos\psi - \upsilon_N \sin\psi \\ \upsilon_E \sin\psi \\ -\upsilon_E \cos\psi \end{Bmatrix} \tag{3.3b}$$

The derivation of the Coriolis force is based on the difference between the total position vector and the Earth-related flow vector: we should consider simultaneously the sphere rotation and relative motions, $r_s^{Total} = r_s + r_s^{Relative}$. A total velocity vector can be presented as follows:

$$\begin{aligned} \left.\frac{d\vec{r}}{dt}\right|_{Total} &= \left(\frac{dr_x}{dt}\vec{i} + \frac{dr_y}{dt}\vec{j} + \frac{dr_z}{dt}\vec{k}\right) + \left(r_x\frac{d}{dt}\vec{i} + r_y\frac{d}{dt}\vec{j} + r_z\frac{d}{dt}\vec{k}\right) \\ \left.\frac{dr_l}{dt}\right|_{Total} &= \frac{dr_l}{dt}i_l + r_l\frac{di_l}{dt} \end{aligned} \tag{3.4a}$$

or we may write

$$\begin{aligned} \left.\frac{d\vec{r}}{dt}\right|_{Total} &= \frac{d\vec{r}}{dt} + \vec{\omega} \times \vec{r} \text{ or } \left.\frac{d\vec{r}}{dt}\right|_{Total} \leftrightarrow \frac{dr_s}{dt} + \varepsilon_{skn}\omega_k r_n \\ \left.\frac{dr_i}{dt}\right|_{Total} &= \frac{dr_i}{dt} + \varepsilon_{ikn}\omega_k r_n \end{aligned} \tag{3.4b}$$

where $\frac{dr_s}{dt}$ means a relative field $\left.\frac{dr_s}{dt}\right|_{\text{Relative}}$.

Consequently, for the second derivatives we obtain

$$\begin{aligned} &\frac{d}{dt}\left[\left.\frac{d\vec{r}}{dt}\right|_{Total}\right]_{Total} = \frac{d}{dt}\left[\frac{d\vec{r}}{dt} + \vec{\omega}\times\vec{r}\right]_{Total} \rightarrow \\ &\frac{d}{dt}\left[\frac{dr_i}{dt} + \varepsilon_{ikn}\omega_k r_n\right]_{Total} = \frac{d}{dt}\left[\frac{dr_i}{dt} + \varepsilon_{ikn}\omega_k r_n\right] + \varepsilon_{ikn}\omega_k\left[\frac{dr_n}{dt} + \varepsilon_{nls}\omega_l r_s\right] = \\ &\frac{d^2 r_i}{dt^2} + 2\varepsilon_{ikn}\omega_k\frac{dr_n}{dt} + \varepsilon_{ikn}\varepsilon_{nls}\omega_k\omega_l r_s \\ &\text{or} \\ &\frac{d}{dt}\left[\left.\frac{d\vec{r}}{dt}\right|_{Total}\right]_{Total} = \frac{d^2\vec{r}}{dt^2} + 2\vec{\omega}\times\frac{d\vec{r}}{dt} + \vec{\omega}\times\vec{\omega}\times\vec{r} \end{aligned} \tag{3.5a}$$

Otherwise, we write for the Coriolis acceleration and a centripetal acceleration term:

$$\rho \frac{dv_s^{Total}}{dt} = \rho \frac{dv_s}{dt} + 2\rho\varepsilon_{skn}\omega_k v_s + \rho\varepsilon_{skn}\varepsilon_{npm}\omega_k\omega_p r_m \tag{3.5b}$$

These expressions for the total accelerations have been obtained with the help of the reference operators applied for the total time derivatives with reference to the Earth rotation.

We may here put a question whether a very long-lasting action of the Coriolis accelerations might have influenced equatorial differences of the seismic activity through an action on the deep fluid part of the Earth interior.

Similarly to the Coriolis forces related to the Earth rotation we may put a question on a similar mechanism acting due to the rotation of the Earth around the Sun. Similarly to relations (3.5) we may write for the position vector related to the Sun center, $R_s^{Total} = R_s + R_s^{Relative}$, and the total velocity vector related to the Earth motion around the Sun:

$$\begin{aligned}
&\left.\frac{d\vec{R}}{dt}\right|_{Total} = \frac{d\vec{R}}{dt} + \vec{\Omega}\times\vec{R} \rightarrow \left.\frac{dR_n}{dt}\right|_{Total} = \frac{dR_n}{dt} + \varepsilon_{nks}\Omega_k R_s \\
&\frac{d}{dt}\left[\frac{dR_n}{dt} + \varepsilon_{nks}\Omega_k R_s\right]_{Total} = \frac{d}{dt}\left[\frac{dR_n}{dt} + \varepsilon_{nks}\Omega_k R_s\right] + \varepsilon_{nks}\Omega_k\left[\frac{dR_s}{dt} + \varepsilon_{sml}\Omega_m R_l\right] = \\
&\frac{d^2R_n}{dt^2} + 2\varepsilon_{nks}\Omega_k\frac{dR_s}{dt} + \varepsilon_{nks}\varepsilon_{sml}\Omega_k\Omega_m R_l \\
&\frac{d}{dt}\left[\left.\frac{d\vec{R}}{dt}\right|_{Total}\right]_{Total} = \frac{d}{dt}\left[\frac{d\vec{R}}{dt} + \vec{\Omega}\times\vec{R}\right]_{Total}
\end{aligned} \tag{3.6a}$$

and

$$\begin{aligned}
&\frac{d}{dt}\left[\frac{dR_i}{dt} + \varepsilon_{ikn}\Omega_k R_n\right]_{Total} = \frac{d}{dt}\left[\frac{dR_i}{dt} + \varepsilon_{ikn}\Omega_k R_n\right] + \varepsilon_{ikn}\Omega_k\left[\frac{dR_n}{dt} + \varepsilon_{nls}\Omega_l R_s\right] = \\
&\frac{d^2R_i}{dt^2} + 2\varepsilon_{ikn}\Omega_k\frac{dR_n}{dt} + \varepsilon_{ikn}\varepsilon_{nls}\Omega_k\Omega_l R_s \\
&\left.\frac{d\vec{R}}{dt}\right|_{Total} = \sum\frac{dR_s}{dt}\vec{i}_s + \varepsilon_{skn}\frac{d\vec{i}_k}{dt}R_n \ where\ \varepsilon_{skn} \rightarrow \varepsilon_{Rkn} \\
&\frac{d}{dt}\left[\left.\frac{d\vec{R}}{dt}\right|_{Total}\right]_{Total} = \frac{d}{dt}\left[\sum\frac{dR_s}{dt}\vec{i}_s + \varepsilon_{skn}\frac{d\vec{i}_k}{dt}R_n\right]_{Total} = \\
&\frac{d}{dt}\left[\sum\frac{dR_s}{dt}\vec{i}_s + \varepsilon_{skn}\frac{d\vec{i}_k}{dt}R_n\right] + \vec{\Omega}\times\left[\sum\frac{dR_s}{dt}\vec{i}_s + \varepsilon_{skn}\frac{d\vec{i}_k}{dt}R_n\right]
\end{aligned} \tag{3.6b}$$

finally

$$\frac{d}{dt}\left[\left.\frac{d\vec{R}}{dt}\right|_{Total}\right]_{Total} = \frac{d^2\vec{R}}{dt^2} + 2\vec{\Omega}\times\frac{d\vec{R}}{dt} + \vec{\Omega}\times\vec{\Omega}\times\vec{R}$$

$$\rho\frac{d\upsilon_s^{Total}}{dt} = \rho\frac{d\upsilon_s}{dt} + 2\rho\varepsilon_{skn}\Omega_k\Omega_s + \rho\varepsilon_{skn}\varepsilon_{npm}\Omega_k\Omega_p R_m \text{ ; where } \upsilon_s^{Total} = \left.\frac{d\vec{R}}{dt}\right|_{Total} \tag{3.6c}$$

where $\frac{d\vec{R}}{dt}$ means $\left.\frac{d\vec{R}}{dt}\right|_{\text{Relative}}$.

Consequently, for the second derivatives we obtain

$$\frac{d}{dt}\left[\left.\frac{d\vec{R}}{dt}\right|_{Total}\right]_{Total} = \frac{d}{dt}\left[\frac{d\vec{R}}{dt} + \vec{\Omega}\times\vec{R}\right]_{Total} =$$
$$\frac{d}{dt}\left[\frac{d\vec{R}}{dt} + \vec{\Omega}\times\vec{R}\right] + \vec{\Omega}\times\left[\frac{d\vec{R}}{dt} + \vec{\Omega}\times\vec{R}\right] \tag{3.6d}$$

or

$$\frac{d}{dt}\left[\left.\frac{d\vec{R}}{dt}\right|_{Total}\right]_{Total} = \frac{d^2\vec{R}}{dt^2} + 2\vec{\Omega}\times\frac{d\vec{R}}{dt} + \vec{\Omega}\times\vec{\Omega}\times\vec{R}$$

The Sun gravitational force and the accelerations related to the Earth motion must be of course in the equilibrium:

$$F_G = m\frac{GM}{R^2} \rightarrow \frac{GM}{R^2}\frac{\vec{R}}{R} = \frac{d^2\vec{R}}{dt^2} + 2\vec{\Omega}\times\frac{d\vec{R}}{dt} + \vec{\Omega}\times\vec{\Omega}\times\vec{R} = \frac{GM}{R^2}\frac{\vec{R}}{R} \tag{3.7}$$

For related accelerations we obtain

$$\frac{GM}{R^2}\frac{R_s}{R} = \frac{d\upsilon_s^{Total}}{dt} = \rho\frac{d\upsilon_s}{dt} + 2\rho\varepsilon_{skn}\Omega_k\Omega_s + \rho\varepsilon_{skn}\varepsilon_{npm}\Omega_k\Omega_p R_m \tag{3.8}$$

Thus, we may deal both with the float transports and reference transports related to the different acceleration reference systems; we may write some combination of these float and reference systems:

$$\rho\frac{d\upsilon_i}{dt} + \rho\sum_s \upsilon_s\frac{\partial\upsilon_i}{\partial x_s} + \rho\left(\sum_s \bar{\upsilon}_s\frac{\partial\upsilon_i}{\partial x_s} + \sum_s \tilde{\upsilon}_s\frac{\partial\upsilon_i}{\partial x_s}\right) = \eta\sum_s\frac{\partial^2\upsilon_i}{\partial x_s^2} + F_i$$
$$\rho\frac{d\bar{\upsilon}_n}{dt} + \rho\sum_s \bar{\upsilon}_s\frac{\partial\bar{\upsilon}_n}{\partial x_s} + \rho\left(\sum_s \upsilon_s\frac{\partial\bar{\upsilon}_n}{\partial x_s} + \sum_s \tilde{\upsilon}_s\frac{\partial\upsilon_i}{\partial x_s}\right) = \eta\sum_s\frac{\partial^2\bar{\upsilon}_n}{\partial x_s^2} + \bar{F}_n$$
$$\rho\frac{d\tilde{\upsilon}_s}{dt} + 2\rho\varepsilon_{skn}\omega_k\tilde{\upsilon}_n + \rho\varepsilon_{skn}\varepsilon_{npm}\omega_k\omega_p r_m + \rho\sum_k \upsilon_k\frac{\partial\tilde{\upsilon}_s}{\partial x_k} + \rho\sum_s \bar{\upsilon}_s\frac{\partial\tilde{\upsilon}_s}{\partial x_s} = \eta\sum_s\frac{\partial^2\upsilon_i}{\partial x_s^2} + \tilde{F}_i \tag{3.9}$$

Finally, let us return to the float motions. First, to the former considerations (cf., Eqs. 3.1–3.3), and to an action of the Coriolis force (Eq. 3.5a, b), we add some local float transport related to the linear motion, $\sum_k \bar{\upsilon}_k \frac{\partial \upsilon_n}{\partial x_k}$:

$$\rho \frac{d\upsilon_s}{dt} + 2\rho\varepsilon_{skn}\omega_k\upsilon_s + \rho\varepsilon_{skn}\varepsilon_{npm}\omega_k\omega_p r_m + \rho\sum_s \upsilon_s \frac{\partial \upsilon_i}{\partial x_s} + \rho\sum_s \bar{\upsilon}_s \frac{\partial \upsilon_i}{\partial x_s} = \eta\sum_s \frac{\partial^2 \upsilon_i}{\partial x_s^2} + \tilde{F}_i \quad (3.10a)$$

or vortex, $\varepsilon_{ikn}\tilde{\upsilon}_k \frac{\partial \upsilon_n}{\partial x_k}$, transport terms:

$$\rho \frac{d\upsilon_s}{dt} + 2\rho\varepsilon_{skn}\omega_k\upsilon_s + \rho\varepsilon_{skn}\varepsilon_{npm}\omega_k\omega_p r_m + \rho\sum_s \upsilon_s \frac{\partial \upsilon_i}{\partial x_s} + \rho\varepsilon_{ikn}\tilde{\upsilon}_k \frac{\partial \upsilon_n}{\partial x_k} = \eta\sum_s \frac{\partial^2 \upsilon_i}{\partial x_s^2} + \tilde{F}_i \quad (3.10b)$$

We thus learned that when dealing with more than one variable motions, e.g., two float transports, $\tilde{\upsilon}_k$ and $\bar{\upsilon}_k$, we should consider two combined relations.

Analysing this example we should underline again the powerful possibilities related to these float transport motions, which in a particular case can join the influences of a solid float transport (e.g., related to the Earth rotation) with a fluid transport (e.g., related to the river flow motions).

Finally, we might consider a number of the mutually related transports, υ_s^τ with s = (1,2,3) and float transport motions, υ_s^τ with $\tau = (1, \ldots\ldots N)$:

$$\rho \frac{d\upsilon_i^\sigma}{dt} + \rho\sum_s \upsilon_s^\sigma \frac{\partial \upsilon_i^\sigma}{\partial x_s} + \rho\sum_{s=1}^{s=3}\sum_{\tau=1}^{\tau=N} \upsilon_s^\tau \frac{\partial \upsilon_i^\sigma}{\partial x_s} = \eta\sum_s \frac{\partial^2 \upsilon_i^\sigma}{\partial x_s^2} + F_i^\sigma \quad (3.11)$$

3.3 Motions in Solids

We introduce a new element into the motions appearing in solids, namely, the molecular transport; this element corresponds to the molecular strains introduced into fluid theory. The molecular transport might play an essential role in the fracture processes.

In the Asymmetric Continuum Theory the deformation fields can be related to the fracture processes; due to some displacement motions in a source, these displacements may belong to an individual process or to some correlated fracture events; hence, the appearance of phase shifts. A point fracture event could appear as related either to a confining load, or/and to the shear and rotation fracture process. However, due to the existence of defects (see Chap. 7) the fault plane mechanism and related emitted waves may not directly correspond to the applied or existing stresses. Moreover, an appearance of some molecular transport motions might also interact here; we may note that the molecular transport could interact with the molecular momentum flux. All these processes and correlated events are mutually related in the release-rebound source mechanism. It is important to note that the existing world-wide seismological network cannot record such strains and molecular motions.

The strains and rotation fields (cf., Asymmetric Strain Theory; Teisseyre 2009, 2011) can be not sufficient to describe the micro-fracture source processes; traditionally, we can present any mechanical events by displacements or displacement derivatives related to the Newton motion equation or to its derivatives. We should take into account that the different, independently released deformations lead, in a final stage, to some macroscopic slip motion; therefore, it may be reasonable to introduce in the premonitory micro-processes, in addition to the Newtonian strain and displacement motion equations, some molecular transport motion. We may follow an analogy to the Asymmetric Molecular Strain Theory (cf., Teisseyre 2010) in which, besides the Navier–Stokes transport relations, we have introduced the asymmetric molecular strains which obey the Newton motion equations for the displacement derivatives (deformations).

In solids, instead of the Navier–Stokes transport of molecular nature, we propose its space–time reverse relation. To these considerations we also include the release-rebound interaction processes.

We will once more briefly repeat, after Chap. 2, the main elements of the Asymmetric Theory in which all the independent motions are described by separate motion relations; the asymmetric strain fields, $E_{ik} = E_{(ik)} + E_{[ik]}$, can be defined as follows:

$$\bar{E} = \frac{1}{3}\sum_{s=1}^{3} E_{(ss)} = \frac{1}{3}\sum_{s=1}^{3} \frac{\partial \bar{u}_s}{\partial x_s};\ \hat{E}_{(ik)} = \frac{1}{2}\left(\frac{\partial \hat{u}_k}{\partial x_i} + \frac{\partial \hat{u}_i}{\partial x_k}\right) - \delta_{ik}\frac{1}{3}\sum_{s=1}^{3}\frac{\partial \hat{u}_s}{\partial x_s}$$
$$E_{[ik]} \equiv \omega_{ik} = \frac{1}{2}\left(\frac{\partial \breve{u}_k}{\partial x_i} - \frac{\partial \breve{u}_i}{\partial x_k}\right) \tag{3.12a}$$

We postulate that all these fields have independent balance relations and that these fields might be either independently released in the source processes, or directly correlated, possibly with some phase shifts:

$$\bar{u}_s = \xi^0 u_s,\ \hat{u}_s = e^0 u_s,\ \breve{u}_s = \chi^0 u_s;\ \left\{\xi^0, e^0, \chi^0\right\} = \{0, \pm 1, \pm \mathrm{i}\} \tag{3.12b}$$

In a similar way, for the total stresses we can write (Teisseyre 2009, 2011):

$$S_{kl} = S_{(kl)} + S_{[kl]} = \delta_{kl}\bar{S} + \hat{S}_{(kl)} + S_{[kl]};$$
$$\hat{S}_{(kl)} = S_{(kl)} - \delta_{ik}\frac{1}{3}\sum_{s=1}^{3} S_{(ss)};\bar{S} = \frac{1}{3}\sum_{s=1}^{3} S_{(ss)} \tag{3.13}$$

The constitutive laws may be taken as follows:

$$S_{(ik)} = 2\mu E_{(ik)} + \lambda\delta_{ik}\sum_{s=1}^{3} E_{(ss)};\ S_{[ik]} = 2\mu E_{[ik]}$$
$$\hat{S}_{(ik)} = 2\mu\hat{E}_{(ik)},\ \breve{S}_{[ik]} = 2\mu\breve{E}_{[ik]},\ \bar{S}_{(ss)} = (2\mu + 3\lambda)\bar{E}_{(ss)} \tag{3.14}$$

The motion equations for these fields, $\bar{S}, \hat{S}_{(kl)}, \breve{S}_{[kl]}$ and $\bar{E}, \hat{E}_{(kl)}, \breve{E}_{[kl]}$, follow from the classic formula for deformation $\frac{\partial u_i}{\partial x_n} = D_{ni}$:

$$\mu\sum\frac{\partial^2}{\partial x_s\partial x_s}\frac{\partial u_n}{\partial x_l} - \rho\frac{\partial^2}{\partial t^2}\frac{\partial u_n}{\partial x_l} = -(\lambda + \mu)\frac{\partial^2}{\partial x_n\partial x_l}\sum\frac{\partial u_s}{\partial x_s} + \frac{\partial F_n}{\partial x_l} \tag{3.15}$$

For the total strains, $\bar{E} = \frac{1}{3}\sum_{s=1}^{3} E_{(ss)}$, we obtain

$$(\lambda + 2\mu)\sum_s\frac{\partial^2\bar{E}}{\partial x_s\partial x_s} - \rho\frac{\partial^2}{\partial t^2}\bar{E} = 0 \tag{3.16a}$$

However, we will confine our considerations only to the wave equations for the deviatoric shear and rotation strains, $\hat{S}_{(kl)}, \breve{S}_{[kl]}$ and $\hat{E}_{(kl)}, \breve{E}_{[kl]}$, under a constant pressure load (cf., Eq. 3.16a):

$$\mu\sum_s\frac{\partial^2\hat{E}_{(ik)}}{\partial x_s\partial x_s} - \rho\frac{\partial^2\hat{E}_{(ik)}}{\partial t^2} = 0$$
$$\mu\sum_s\frac{\partial^2\breve{E}_{[ik]}}{\partial x_s\partial x_s} - \rho\frac{\partial^2\breve{E}_{[ik]}}{\partial t^2} = 0 \tag{3.16b}$$

These shear and rotation strain waves may exist independently of any real displacements; that is, the related displacements, cf., Eq. 3.14, formally describing these strain fields, may not exist in reality.

Besides the independent strain fields we assume an appearance of the molecular transport, as governed by the molecular micro-motions (an idea related to the Navier–Stokes transport, $\upsilon_s = \partial u_s/\partial t$, with interchanged space and time):

$$w_s = \frac{\partial\upsilon_s}{\partial x_s} \to \left\{w_1 = \frac{\partial\upsilon_1}{\partial x_1}, w_2 = \frac{\partial\upsilon_2}{\partial x_2}, w_3 = \frac{\partial\upsilon_3}{\partial x_3}\right\} \tag{3.17}$$

where for solids we assume $\upsilon_s = 0$; however, for some special cases or materials, this condition should be rejected, e.g., for glaciers (see the end of this chapter).

Thus, we propose a new governing relation for these transport micro-motions as governed by the molecular micro-motion equations; namely, we propose to use relations similar to the Navier–Stokes ones, but with a time transport, $\tilde{\tau}(\rho w_s)\frac{\partial w_s}{\partial t}$ (instead of space transport, we introduce here the time related transport):

$$\frac{d(\rho w_s)}{dt} = w_s\frac{\partial \rho}{\partial t} + \rho\frac{\partial w_s}{\partial t} + \tilde{\tau}\rho w_s\frac{\partial w_s}{\partial t} = \tilde{\eta}\sum\frac{\partial^2 w_s}{\partial x_k \partial x_k} \tag{3.18a}$$

or we obtain for a constant density in this micro-range:

$$\rho\frac{\partial w_s}{\partial t} + \tilde{\tau}\rho w_s\frac{\partial w_s}{\partial t} = \tilde{\eta}\sum\frac{\partial^2 w_s}{\partial x_k \partial x_k} \tag{3.18b}$$

where $\tilde{\tau}$ is a relaxation time and $\tilde{\eta}$ means a dynamic micro-viscosity (the new material constants).

This relation explains a possible wave motion of this field; we note that an intensity of such motions could increase in seismically active regions.

A more general approach to the transport processes has been presented by Landau and Lifshitz (1959), new edition translated into Polish as Landau and Lifszyc (2009); first we recall the definition of the momentum flux related to the molecular strains, $\tilde{S}$, in fluids,

$$\tilde{T}_{kn} = -\tilde{S}_{kn} + \rho\upsilon_k\upsilon_n\,,\ \tilde{S}_{(ij)} = \eta\left(\frac{\partial\upsilon_j}{\partial x_i} + \frac{\partial\upsilon_i}{\partial x_j} - \frac{2\delta_{ij}}{3}\sum_s\frac{\partial\upsilon_s}{\partial x_s}\right) + \varepsilon\delta_{ij}\sum_s\frac{\partial\upsilon_s}{\partial x_s} \tag{3.19}$$

and the Landau-Lifshitz relation at a constant density:

$$\rho\frac{\partial\upsilon_n}{\partial t} + \sum_k\frac{\partial\tilde{T}_{kn}}{\partial x_k} = 0 \tag{3.20a}$$

This relation can be supplemented by the antisymmetric stress rates (Teisseyre 2010):

$$\begin{gathered}\tilde{T}_{kn} = \tilde{T}_{(kn)} + \tilde{T}_{[kn]} = -\tilde{S}_{kn} + \rho\upsilon_k\upsilon_n,\ \tilde{T}_{(kn)} = -\tilde{S}_{(kn)} + \rho\upsilon_k\upsilon_n,\ \tilde{T}_{[kn]} = -\tilde{S}_{[kn]},\ \tilde{S}_{kn} = \tilde{S}_{(kn)} + \tilde{S}_{[kn]}\\ \tilde{S}_{(kn)} = \eta\left(\frac{\partial\upsilon_n}{\partial x_k} + \frac{\partial\upsilon_k}{\partial x_n} - \frac{2\delta_{kn}}{3}\sum_s\frac{\partial\upsilon_s}{\partial x_s}\right) + \varepsilon\delta_{kn}\sum_s\frac{\partial\upsilon_s}{\partial x_s},\ \tilde{S}_{[kn]} = \eta\left(\frac{\partial\upsilon_n}{\partial x_k} - \frac{\partial\upsilon_k}{\partial x_n}\right)\end{gathered} \tag{3.20b}$$

According to Eq. (3.17) we redefine these stress micro-fields and related equations; in a similarity to the stress–strain relations (3.11–3.17) we propose for the micro-strains at vanishing total micro-strains the following equation:

$$\ddot{E} = \frac{1}{3}\sum_{s=1}^{3}\frac{\partial w_s}{\partial x_s} = 0\,;\ \ddot{E}_{(ik)} = \frac{1}{2}\left(\frac{\partial w_k}{\partial x_i} + \frac{\partial w_i}{\partial x_k}\right),\ \ddot{E}_{[ik]} = \frac{1}{2}\left(\frac{\partial w_k}{\partial x_i} - \frac{\partial w_i}{\partial x_k}\right) \tag{3.21a}$$

and for the micro-stresses we write:

$$\ddot{S}_{(ik)} = 2\mu\ddot{E}_{(ik)};\ \ddot{S}_{[ik]} = 2\mu\ddot{E}_{[ik]} \tag{3.21b}$$

$$\ddot{S}_{(kn)} = \ddot{\eta}\left(\frac{\partial w_n}{\partial x_k} + \frac{\partial w_k}{\partial x_n}\right), \ddot{S}_{[kn]} = \ddot{\eta}\left(\frac{\partial w_n}{\partial x_k} - \frac{\partial w_k}{\partial x_n}\right) \text{ and } \ddot{S}_{kn} = 2\ddot{\eta}\frac{\partial w_n}{\partial x_k} \tag{3.21c}$$

where the new parameter introduced, $\ddot{\eta}$, presents the material parameter related to a fracture recess.

Now, we can write a general relation for the asymmetric momentum flux with a molecular transport and constant density:

$$\ddot{T}_{kn} = \ddot{T}_{(kn)} + \ddot{T}_{[kn]} = -\ddot{S}_{kn} + \rho w_k w_n = -2\ddot{\eta}\frac{\partial w_n}{\partial x_k} + \rho w_k w_n \tag{3.21d}$$

The obtained relations for micro-transport constitute the basic relations for the micro-transport motions, especially in a kind of fracture recess; we arrive at the new Landau-Lifshitz-like relation:

$$\rho\frac{\partial v_n}{\partial t} + \sum_k \frac{\partial T_{kn}}{\partial x_k} = F_n \rightarrow \rho\frac{\partial w_n}{\partial t} + \sum_k \frac{\partial \ddot{T}_{kn}}{\partial x_k} = \ddot{F}_n \tag{3.21e}$$

and finally we have:

$$\rho\frac{\partial w_n}{\partial t} - 2\ddot{\eta}\sum_k \frac{\partial^2 w_n}{\partial x_k \partial x_k} = \ddot{F}_n \tag{3.21f}$$

The considered micro-motions could relate also to a rotation transport field, $\tilde{w}$, as defined by the components of micro-rotation velocities $\tilde{r}, r\tilde{\omega}, \tilde{z}$ (cf., Eq. 3.17):

$$w_r = \frac{\partial \tilde{r}}{\partial r}, \; w_\varphi = \frac{\partial \tilde{\omega}}{\partial \varphi}, \; w_r = \frac{\partial \tilde{z}}{\partial z} \tag{3.22a}$$

where the rotation-related velocities vanish as well $(\tilde{r} = 0, \tilde{\omega} = 0, \tilde{z} = 0)$.

Accordingly, the micro-transport relations (3.21f) become expressed by the flow relations:

$$\rho\frac{\partial w_r}{\partial t} - 2\ddot{\eta}\Delta w_r = \ddot{F}_r, \; \rho\frac{\partial w_\varphi}{\partial t} - 2\ddot{\eta}\Delta w_\varphi = \ddot{F}_\varphi, \; \rho\frac{\partial w_z}{\partial t} - 2\ddot{\eta}\Delta w_z = \ddot{F}_z \tag{3.22b}$$

We have arrived at a definition for the asymmetric micro-momentum flux with the adequate material micro-constants. Here, in Eqs. (3.21) and (3.22), we have added a possible force rate (on the right-hand side); such force rates can be related to an influence of the elastic strain waves (cf., Eq. 3.16); we should note that these strains related to the actually arriving waves may be an important element in the micro-processes and hence could excite some activity in the seismic regions. We notice that when considering the terms $\Delta w_r, \Delta w_\varphi, \Delta w_z$ with $w_r(r, \varphi, z) = w_r(r)w_r(\varphi)w_r(z)$, $w_\varphi(r, \varphi, z) = w_\varphi(r)w_\varphi(\varphi)w_\varphi(z)_z$, $w_z(r, \varphi, z) = w_z(r)w_z(\varphi)w_z(z)_z$, we should be careful when searching for the related wave solutions for the terms $w_r(r)$,$w_\varphi(r)$,$w_z(r)$ which should be the non-harmonic solutions.

It is worth noting here that the definitions for micro-rotation velocities are different from those following from a simple coordinate transformation ($r\tilde{\omega}$ instead of $\tilde{r\omega}$) bringing the important physical consequences to our considerations.

We should also underline that the velocity rotation transports, $\tilde{r}, r\tilde{\omega}, \tilde{z}$, are introduced instead of the classical approach to vorticity as rotation displacement fields (cf., Teisseyre 2009). Such flows become effectively important for the Reynolds numbers above the critical value.

Finally, we arrive at a fracture process; the micro-fracture relations (3.21, 3.22), especially important for fracture recess domains, can be used to describe the fracture slip process.

In particular, for a fracture process related to micro-transport w_x, on a plane (y,z), we may consider the partial differentiations of these micro-transport relations inside a fracture recess domain elongated in the y and z directions:

$$\rho\frac{\partial}{\partial t}\frac{\partial^2 w_x}{\partial y\partial z} - 2\eta\Delta\frac{\partial^2 w_x}{\partial y\partial z} = \frac{\partial^2 \ddot{F}_x}{\partial y\partial z} \rightarrow \rho\frac{\partial w_x}{\partial t} - 2\eta\Delta w_x = \ddot{F}_x \tag{3.22c}$$

Thus, we obtain the equation for a related slip.

These considerations present a theoretical study related to the molecular-transport, w_x, w_y, w_z, in a solid material under stress vibrations caused by the seismic elastic waves; a micro-transport means here a time rate of velocity transport, where we assume that these velocities vanish: $\upsilon_1 = 0$, $\upsilon_2 = 0$, $\upsilon_3 = 0$, while: $\frac{\partial \upsilon_s}{\partial x_s} \rightarrow \{w_1, w_2, w_3\}$.

In a similar way to (3.17), we may define the molecular displacements:

$$w_s^u = \frac{\partial u_s}{\partial x_s} \rightarrow \left\{ w_1^u = \frac{\partial u_1}{\partial x_1}, w_2^u = \frac{\partial u_2}{\partial x_2}, w_3^u = \frac{\partial u_3}{\partial x_3} \right\} \tag{3.23a}$$

leading to the related equation:

$$\rho\frac{\partial w_n^u}{\partial t} - 2\mu\sum_k \frac{\partial^2 w_n^u}{\partial x_k \partial x_k} = \ddot{F}_n^u \tag{3.23b}$$

We have introduced new material constants related to those hypothetical micro-transport motions. No direct experiments confirm the existence of the micro-transport in solids; however, the complicated processes in a material under high stress loads well confirm this theoretical approach.

3.4 Motions in Fluids

In our former papers (Teisseyre 2009, 2011, 2012) related to a solid state we have included into the theory the antisymmetric strains; the experimental seismological observations proved the existence of rotation motions, that is, the antisymmetric strains. The problems were discussed at two international conferences related to "rotation seismology", one in Menlo Park USA, 2008 (cf., Lee et al. 2009), and the other in Prague 2009 (cf., Teisseyre 2011). These data and related theoretical considerations give grounds for including the antisymmetric molecular strains in the fluid theory; some consideration on the theory of vortex motions (Teisseyre 2009) supports this approach. In a discussion on basic motions and relations in fluids we should remember the main elements of the Asymmetric Continuum Theory for solids (Teisseyre 2009, 2011, 2012).

In fluids with a variable density we have the displacement velocities, υ, presenting the real fluid transport motions (such transport motions might form also some macro-strains) and an asymmetry in a fluid continuum relates mainly to the molecular counterpart of the strains.

For the molecular symmetric stresses we can write after Landau and Lifschitz (1959); new edition translated into Polish as Landau and Lifszyc (2009):

$$\begin{aligned} \Xi_{nk} &= 2\eta\left(\tilde{E}_{(ij)} - \frac{\delta_{ij}}{3}\sum_s \frac{\partial \tilde{\upsilon}_s}{\partial x_s}\right) + \xi\delta_{ij}\sum_s \frac{\partial \tilde{\upsilon}_s}{\partial x_s} \rightarrow \\ \Xi_{nk} &= \eta\left(\frac{\partial \tilde{\upsilon}_j}{\partial x_i} + \frac{\partial \tilde{\upsilon}_i}{\partial x_j} - \frac{2\delta_{ij}}{3}\sum_s \frac{\partial \tilde{\upsilon}_s}{\partial x_s}\right) + \xi\delta_{ij}\sum_s \frac{\partial \tilde{\upsilon}_s}{\partial x_s} \end{aligned} \tag{3.24a}$$

After Landau and Lifshitz, the Euler equation for viscous fluid is supplemented by a tensor Ξ_{nk} which should depend only on first derivatives of flow velocities, that is, on $\frac{\partial \tilde{\upsilon}_j}{\partial x_i}$ and $\delta_{ij}\sum\limits_s \frac{\partial \tilde{\upsilon}_s}{\partial x_s}$; this will permit the fluid properties to become dependent only on the scalars, the viscous constants, η and ξ:

$$\begin{aligned} \rho\frac{d\upsilon_n}{dt} &= -\frac{\partial p}{\partial x_n} + \sum_k \frac{\partial \Xi_{nk}}{\partial x_k} - \sum_k \frac{\partial(\rho\upsilon_n\upsilon_k)}{\partial x_k}; \\ \Xi_{nk} &= \eta\left\{\frac{\partial \upsilon_n}{\partial x_k} + \frac{\partial \upsilon_k}{\partial x_n} - \frac{2}{3}\delta_{nk}\sum_k \frac{\partial \upsilon_k}{\partial x_k}\right\} + \xi\delta_{nk}\sum_s \frac{\partial \upsilon_s}{\partial x_s} \end{aligned} \tag{3.24b}$$

where $\sum\limits_k \Xi_{kk} = 3\xi\sum\limits_s \frac{\partial \upsilon_s}{\partial x_s}$.

However, an essential problem appears here: in a classic approach a mass change is presented as:

$$dm = d(\rho\delta V) \rightarrow \rho d(\delta V) \tag{3.25a}$$

which means that the density in an infinitesimal volume, δV, should be constant.

However, for a gas medium this assumption may be false; e.g., for a nuclear explosion we should rather write that an element of mass, dm, related to released energy will influence immediately both the equivalent density, $d\rho$, and a change of a volume element, $d(\delta V)$. In this approach a mass change should not be written as $d'\frac{E}{c^2} = d'm = d'(\rho\delta V)$, but rather as follows:

$$d'\frac{E}{c^2} = d'm \rightarrow d'm = d'(\rho\delta V) = \rho d'(\delta V) \tag{3.25b}$$

where the symbol d' means small elements of mass or volume.

Thus, we may consider the Euler equation:

$$d'm = d'(\rho\delta V) \rightarrow d'm\frac{\partial v_i}{\partial t} = -\frac{\partial p}{\partial x_i} + F_i \rightarrow \rho d'(\delta V)\frac{\partial v_i}{\partial t} = -\frac{\partial p}{\partial x_i} + F_i \tag{3.25c}$$

and when arriving at the derivatives we will obtain:

$$\left(\delta V\frac{\partial \rho}{\partial t} + \rho\frac{\partial \delta V}{\partial t}\right)\frac{\partial v_i}{\partial t} + \rho(\delta V)\frac{\partial^2 v_i}{\partial t^2} = -\frac{\partial^2 p}{\partial t \partial x_i} + \frac{\partial F_i}{\partial t} \tag{3.25d}$$

We have obtained the full transport equation, with the transport fields, the dynamic viscosity, body force rate and with the changes of density.

We should note that the term $\frac{\partial^2 p}{\partial t \partial x_i}$ is related to the acoustic waves, $\frac{\partial p}{\partial t}$, and explain that an acceleration of the flow velocities $\frac{\partial^2 v_i}{\partial t^2}$ in Eq. (3.25d) will appear only in some special cases, e.g., when we deal with a rapid energy release influencing immediately either a density, ρ, or a small volume element, $d'(\delta V)$.

An acoustic effect, related to $\frac{\partial p}{\partial t}$, is associated here with some inner process related to microscopic material behaviour at some rapid process, e.g., a kind of pellicle fracture, or in other words, the destruction of an internal molecular structure.

Further we may note that the relatively very low acoustic wave velocity in liquids is related to pressure and density variations, $c = \sqrt{\partial p/\partial \rho}$. However, for a gas we should put $c = \sqrt{\gamma p/\rho}$, where γ is the adiabatic index. These relations are valid when the sound waves are treated as a small perturbation only. The considerations on acoustic waves will not be continued here.

Returning to the Navier–Stokes relations (3.24) and with (3.25) we should make a replacement: $\Xi_{nk} \rightarrow \tilde{S}_{nk}$, to obtain:

$$\rho\frac{dv_i}{dt} = \rho\frac{\partial v_i}{\partial t} + \rho\sum_s v_s\frac{\partial v_i}{\partial x_s} = \sum_k\frac{\partial \tilde{S}_{ik}}{\partial x_k} - \frac{\partial p}{\partial x_i} + F_i \quad \text{at} \quad \rho v_i v_k = 0 \tag{3.26a}$$

In this relation we take into account that density changes might appear very quickly (e.g., due to radiation), thus we assume that the time changes of momentum are equal to forces.

The important questions appear when returning from to the Euler equation for viscouse fluid to the Landau and Lifshitz (1959, 2009) equations. While considering the time changes of a full set of the applied stresses we obtain only those related to a pressure field (cf., 3.24b):

$$\rho\frac{\partial\upsilon_i}{\partial t} + \rho\sum_s \upsilon_s\frac{\partial\upsilon_i}{\partial x_s} = -\frac{\partial p}{\partial x_i} + \eta\left\{\sum_k \frac{\partial^2\upsilon_i}{\partial x_k\partial x_k} + \frac{\partial}{\partial x_i}\left(\sum_k\frac{\partial\upsilon_k}{\partial x_k} - \frac{2}{3}\sum_l\frac{\partial\upsilon_l}{\partial x_l}\right)\right\}$$
$$+ \xi\frac{\partial}{\partial x_i}\sum_s\frac{\partial\upsilon_s}{\partial x_s} + F_i \tag{3.26b}$$

where we have included the corrected asymmetric forms, $\tilde{S}_{ij} = \tilde{S}_{(ij)} + \tilde{S}_{[kn]} = 2\eta\left(\frac{\partial\upsilon_j}{\partial x_i} - \frac{\delta_{ij}}{3}\sum_s\frac{\partial\upsilon_s}{\partial x_s}\right) + \xi\delta_{ij}\sum_s\frac{\partial\upsilon_s}{\partial x_s}$ (cf., Eq. 3.20b); in a general case, the parameters η and ξ might depend on the pressure and temperature.

However, when we also introduce the asymmetric stress/strain rates then we should put a slightly different equation because we have: $\tilde{S}_{ij} \to \tilde{S}_{(ij)} + \tilde{S}_{[ij]} = 2\eta\left(\frac{\partial\tilde{\upsilon}_j}{\partial x_i} - \frac{\delta_{ij}}{3}\sum_s\frac{\partial\tilde{\upsilon}_s}{\partial x_s}\right) + \xi\delta_{ij}\sum_s\frac{\partial\tilde{\upsilon}_s}{\partial x_s}$ with $\sum_k\tilde{S}_{kk} = 3\xi\sum_s\frac{\partial\upsilon_s}{\partial x_s}$ (cf., 3.24b):

$$\upsilon_i\frac{\partial\rho}{\partial t} + \rho\frac{\partial\upsilon_i}{\partial t} + \rho\sum_s\upsilon_s\frac{\partial\upsilon_i}{\partial x_s} = -\frac{\partial p}{\partial x_n} + 2\eta\left(\sum_k\frac{\partial^2\upsilon_i}{\partial x_k\partial x_k} - \frac{1}{3}\frac{\partial}{\partial x_i}\sum_s\frac{\partial\upsilon_s}{\partial x_s}\right)$$
$$+ \xi\frac{\partial}{\partial x_i}\sum_s\frac{\partial\upsilon_s}{\partial x_s} + F_i \tag{3.27}$$

Returning to the transport relation (3.26a) we may propose an essential change:

$$\frac{d(\rho\upsilon_i)}{\mathrm{d}t} = \upsilon_i\frac{\partial\rho}{\partial t} + \rho\frac{\partial\upsilon_i}{\partial t} + \rho\sum_s\upsilon_s\frac{\partial\upsilon_i}{\partial x_s} = \sum_k\frac{\partial\tilde{S}_{ik}}{\partial x_k} - \frac{\partial p}{\partial x_i} + F_i \text{ at } \rho\upsilon_i\upsilon_k = 0 \tag{3.28a}$$

The density changes might appear very quickly (e.g., due to radiation). A counterpart to this relation is the mass conservation law:

$$\frac{d\rho}{dt} = \frac{\partial\rho}{\partial t} + \sum_s\frac{\partial(\rho\upsilon_s)}{\partial x_s} = 0 \tag{3.28b}$$

while a combination of these two relations leads to

$$-\upsilon_i\sum_s\frac{\partial(\rho\upsilon_s)}{\partial x_s} + \rho\frac{\partial\upsilon_i}{\partial t} + \rho\sum_s\upsilon_s\frac{\partial\upsilon_i}{\partial x_s} = \sum_k\frac{\partial\tilde{S}_{ik}}{\partial x_k} - \frac{\partial p}{\partial x_i} + F_i \tag{3.28c}$$

Finally we obtain a correlated set of equations:

$$\begin{aligned}&\rho\frac{\partial\upsilon_i}{\partial t} - \upsilon_i\sum_s\upsilon_s\frac{\partial\rho}{\partial x_s} - \rho\upsilon_i\sum_s\frac{\partial\upsilon_s}{\partial x_s} + \rho\sum_s\upsilon_s\frac{\partial\upsilon_i}{\partial x_s} = \sum_k\frac{\partial\tilde{S}_{ik}}{\partial x_k} - \frac{\partial p}{\partial x_i} +\\ &F\frac{\partial\rho}{\partial t} + \sum_s\upsilon_s\frac{\partial\rho}{\partial x_s} + \rho\sum_s\frac{\partial\upsilon_s}{\partial x_s} = 0\end{aligned} \tag{3.28d}$$

3.5 Glaciers

Considering the flow and fracture properties in glaciers, we may combine some considerations presented above.

Thus, we may briefly consider fractures in glaciers (see Górski 2004, 2013). This author has discovered another type of fractures, in addition to the typical icequakes. Such special events do not have a sharp beginning, but rather the slowly increasing amplitudes. It seems that such processes, named the ice-vibrations (Górski 2004), can be considered as the flow-fracture events with a mixture of the solid fracture and fluid flow processes. Thus, these ice-vibrations (flow-fracture events) might appear due to the glacier properties; under a high load of the upper part of a glacier, the plastic glacier flow events may appear somewhere in a lower part. A rock background topography beneath a glacier has also some influence on such processes. In this way we may have a mixture of glacier properties with its solid and fluid counterparts.

The related processes may be described by the following combined system of equations; we write after Eq. (3.18a) and with (3.17):

$$w_s \frac{\partial \rho}{\partial t} + \rho \frac{\partial w_s}{\partial t} + \tilde{\tau} \rho w_s \frac{\partial w_s}{\partial t} = \tilde{\eta} \sum \frac{\partial^2 w_s}{\partial x_k \partial x_k} \tag{3.29a}$$

where $w_k = \frac{\partial \upsilon_k}{\partial x_k}$ (cf., 3.17), and according to Eq. (3.28a) we have:

$$\frac{d\rho}{dt} = \frac{\partial \rho}{\partial t} + \rho \sum_s w_s + \sum_s \upsilon_s \frac{\partial \rho}{\partial x_s} = 0 \tag{3.29b}$$

From these two relations we may eliminate $\frac{\partial \rho}{\partial t}$, arriving at:

$$\rho(1 + \tilde{\tau} w_s) \frac{\partial w_s}{\partial t} - \rho w_s \sum_k w_k - w_s \sum_k \upsilon_k \frac{\partial \rho}{\partial x_k} = \tilde{\eta} \sum \frac{\partial^2 w_s}{\partial x_k \partial x_k} \tag{3.29c}$$

And finally we arrive at the system (3.29c) and (3.27) with the six unknown fields, υ_k and w_s, and the properly assumed external relations for $\frac{\partial \rho}{\partial x_s}$ and $\frac{\partial p}{\partial x_i}$:

$$\rho(1 + \tilde{\tau} w_s) \frac{\partial w_s}{\partial t} - \rho w_s \sum_k w_k - w_s \sum_k \upsilon_k \frac{\partial \rho}{\partial x_k} = \tilde{\eta} \sum \frac{\partial^2 w_s}{\partial x_k \partial x_k} \tag{3.30a}$$

$$\begin{aligned} &\rho \frac{\partial \upsilon_i}{\partial t} - \upsilon_i \rho \sum_s w_s - \upsilon_i \sum_s \upsilon_s \frac{\partial \rho}{\partial x_s} + \rho \sum_s \upsilon_s \frac{\partial \upsilon_i}{\partial x_s} \\ &= 2\eta \sum_k \frac{\partial^2 \upsilon_i}{\partial x_k \partial x_k} + \left(\xi - \frac{2\eta}{3} \right) \frac{\partial}{\partial x_i} \sum_s \frac{\partial \upsilon_s}{\partial x_s} - \frac{\partial p}{\partial x_i} + F_i \end{aligned} \tag{3.30b}$$

The external forces are related to a vertical gravity component dependent on glacier thickness and a depth of the considered event and to the flow forces caused by the glacier background topography.

We may shortly considered a very simple case, where, in approximation, we take the system of the *HLZ* directions, where *H* is the glacier flow direction, *L* its perpendicular and *Z* vertical direction assuming that the typical fracture processes extend along the vertical planes, *HZ*. We assume a constant glacier flow, $\upsilon_i \to \upsilon_H$, on the plane *HZ*, and that the flow-fracture events (ice vibrations), $w_s \to w_H$, may appear along a horizontal plane, *HL*. From the system (3.30) we obtain along the *HZ* plane:

$$\begin{aligned} \rho(1+\tilde{\tau}w_H)\frac{\partial w_H}{\partial t} - \rho w_H^2 - w_H\upsilon_H\frac{\partial \rho}{\partial z} &= \tilde{\eta}\frac{\partial^2 w_H}{\partial x_k \partial x_k} \quad \text{(a)} \\ \text{and} \quad -\upsilon_H\rho\bar{w}_H - \upsilon_H^2\frac{\partial \rho}{\partial z} &= -\frac{\partial p}{\partial z} + \rho g \quad \text{(b)} \end{aligned} \tag{3.31a}$$

where in the second relation, (b), the symbol $\bar{w}_H$ means a mean value of the w_H oscillations.

For the first relation, (a), we assume the small oscillation amplitudes, $\tilde{\tau}w_H\frac{\partial w_H}{\partial t} \approx 0$ and $w_H^2 \approx 0$; therefore, we obtain:

$$\rho\frac{\partial w_H}{\partial t} - w_H\upsilon_L\frac{\partial \rho}{\partial z} = \tilde{\eta}\frac{\partial^2 w_H}{\partial x_H \partial x_H} \tag{3.31b}$$

leading to the simple solution for the flow-fracture process (ice vibration):

$$w_H = w\exp(-\alpha t + \beta x_H)\sin(\gamma x_H - \varsigma t) \tag{3.31c}$$

where we should assume $\alpha t \geq \beta x_H$; $\alpha t \geq 0$, $\beta x_H \geq 0$.

We add that the following conditions should be fulfilled:

$$\begin{aligned} &-\alpha\rho w\exp(\beta x_H - \alpha t)\sin(\gamma x_H - \varsigma t) - \varsigma w\exp(\beta x_H - \alpha t)\cos(\gamma x_H - \varsigma t) \\ &-\upsilon_H\frac{\partial \rho}{\partial z}w\exp(\beta x_H - \alpha t)\sin(\gamma x_H - \varsigma t) = \tilde{\eta}w\beta^2\exp(\beta x_H - \alpha t)\sin(\gamma x_H - \varsigma t) + \\ &2\tilde{\eta}w\beta\gamma\exp(\beta x_H - \alpha t)\cos(\gamma x_H - \varsigma t) - \tilde{\eta}w\gamma^2\exp(\beta x_H - \alpha t)\sin(\gamma x_H - \varsigma t) \end{aligned} \tag{3.31d}$$

which become reduced to the conditions:

$$\alpha\rho + \upsilon_H\frac{\partial \rho}{\partial z} + \tilde{\eta}\beta^2 - \tilde{\eta}\gamma^2 = 0 \text{ and } \varsigma + 2\tilde{\eta}\beta\gamma = 0 \tag{3.31e}$$

The desired solution for the flow-fracture (ice vibration) event (3.31c) remains for $\alpha t \geq \beta x_H$.

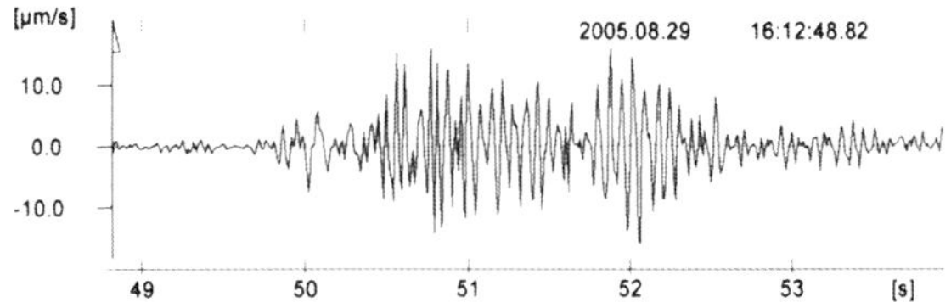

Fig. 3.1 An example of ice vibration record presenting a flow-fracture process

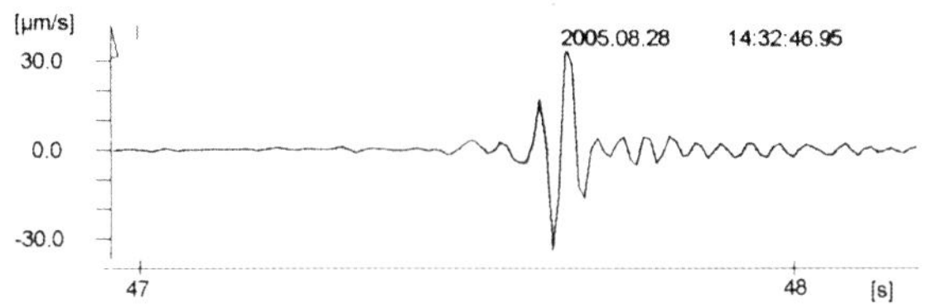

Fig. 3.2 An example of a typical icequake recording

In Figs. 3.1 and 3.2 we present examples of independent ice vibration and icequake records. These events were recorded at the Polish Polar Station in Horsund, Spitsbergen, in two consecutive days in August 2005 (Górski 2013).

References

Górski M (2004) Predominant frequencies in the spectrum of ice-vibration events. Acta Geophysica Pol V 24(4):457–464

Górski M (2013) Seismic events in glaciers. Series: Geoplanet. Earth Planet Sci Series, Springer, Berlin Heidelberg. doi:10.1007/978-3-642-31851-1

Landau LD, Lifschitz JM (1959) Fluid mechanics (Theoretical Physics, v. 6), (translated from Russian by J.B. Sykes and W.H. Reid). Pergamon Press, London, p 536

Landau LD, Lifszyc JM (2009) Hydrodynamika (Hydrodynamics) Wydawnictwo Naukowe PWN Warszawa (Polish translation), p 671

Lee WHK, Celebi M, Todorovska MI, Igel H (2009) Introduction to special issue on rotational seismology and engineering applications. Bull Seismol Soc Am 62(2B):945–957

Teisseyre R (2009) Tutorial on new development in physics of rotation motions. Bull Seismol Soc Am 99(2B):1028–1039

Teisseyre R (2010) Fluid theory with asymmetric molecular stresses: difference between vorticity and spin equations. Acta Geophys 58(6):1056–1071

Teisseyre R (2011) Why rotation seismology: confrontation between classic and asymmetric theories. Bull Seismol Soc Am 101(4):1683–1691

Teisseyre R (2012) Rotation and strain seismology. J Seismol 16(4):683–694. doi:10.1007/s10950-012-9287-6

Chapter 4
Vortices and Molecular Fracture Transport

4.1 Introduction

We discuss a new approach to the vortex processes. Instead of the classic approach based on the definition of vorticity, we consider the vortex motion system as defined directly from the transport processes. The transport field related to the vortex motion can be defined basing on the angular molecular moment field related to rotations with vortex axis along the z-line (using here the cylindrical coordinates). Acting with the curl operator on the Navier–Stokes transport equation we arrive at the combined relation for vorticity and the related transport. The related transport trajectories have the six degrees of freedom in space. This new definition is based on an angular moment with a variable arm and the incorporated spin motions. Next, we consider the momentum flux as related to the new asymmetric form and the approach to the molecular stress system following Landau and Lifshitz (1959). This makes it possible to consider a single stationary vortex, near a boundary of the two opposite-direction uniform flows and even a whole row of vortices along that uniform flow. On this basis we present a model of a single turbulence with the opposite-direction stream flows and a vortex system formed when the relative velocity overpasses a laminar range, as some disturbances may appear due to a fluid viscosity. This theoretical approach seems to be adequate to describe a model of turbulence mechanics based on the vortex dynamics; we believe that this new approach opens a way to considering a mechanical model of vortex and turbulence motions at different thermodynamical conditions.

4.2 Vortices and Molecular Fracture Transport

We will now discuss some specific correlated motions and related interactions, especially for the case of vortex processes. In the classic approach, the vortex processes are described using the definition of vorticity ς:

R. Teisseyre and M. Teisseyre-Jeleńska, *Asymmetric Continuum*,
GeoPlanet: Earth and Planetary Sciences, DOI: 10.1007/978-3-642-31860-3_4,

$$\varsigma_k = \varepsilon_{kpi}\frac{\partial \upsilon_i}{\partial x_p} \tag{4.1a}$$

The vortex motions with their structural dimensions in the cylindrical system, r, φ, z, include rotations inside a considered vortex. Acting with the curl operator on the Navier–Stokes transport equation we arrive at the combined relation for vorticity, ς, and for the related transport:

$$\rho\frac{\partial \varsigma_k}{\partial t} + \rho\sum_s \upsilon_s\frac{\partial \varsigma_k}{\partial x_s} + \rho\varepsilon_{kpi}\sum_s \frac{\partial \upsilon_s}{\partial x_p}\frac{\partial \upsilon_i}{\partial x_s} = \eta\sum_s \frac{\partial^2 \varsigma_k}{\partial x_s^2} - \frac{\varepsilon_{kni}\partial F_i}{\partial x_n} \tag{4.1b}$$

while for a float transport vorticity ς we obtain:

$$\rho\frac{\partial \varsigma_z}{\partial t} + \rho\upsilon_r\frac{\partial \varsigma_z}{\partial r} + \rho\upsilon_\varphi\frac{\partial \varsigma_z}{r\partial \varphi} + \rho\upsilon_z\frac{\partial \varsigma_z}{\partial z} + \rho\varepsilon_{zpi}\frac{\partial \upsilon_s}{\partial x_p}\frac{\partial \upsilon_i}{\partial x_s} = \eta\sum_s \frac{\partial^2 \varsigma_z}{\partial x_s^2} - \frac{\varepsilon_{zni}\partial F_i}{\partial x_n} \tag{4.1c}$$

Thus, our aim is to consider the vortex motion system as defined directly from the transport processes. The transport field related to the vortex motion can be defined basing on the angular molecular moment field related to rotations with the vortex axis along the z-line (using the cylindrical coordinates). We may start with the transport motions in the cylindrical system (see Chap. 3, Eq.3.1b):

$$\begin{aligned}
\rho\frac{d\tilde{\upsilon}_r}{dt} &= \rho\frac{\partial \tilde{\upsilon}_r}{\partial t} + \rho\tilde{\upsilon}_r\frac{\partial \tilde{\upsilon}_r}{\partial r} + \rho\tilde{\upsilon}_\varphi\frac{\partial \tilde{\upsilon}_r}{r\partial \varphi} + \rho\tilde{\upsilon}_z\frac{\partial \tilde{\upsilon}_r}{\partial z} + \rho\sum_s \bar{\upsilon}_s\frac{\partial \tilde{\upsilon}_r}{\partial x_s} = \eta\Delta\tilde{\upsilon}_r + \tilde{F}_r \\
\rho\frac{d\tilde{\upsilon}_\varphi}{dt} &= \rho\frac{\partial \tilde{\upsilon}_\varphi}{\partial t} + \rho\tilde{\upsilon}_\varphi\frac{\partial \tilde{\upsilon}_\varphi}{\partial r} + \rho\tilde{\upsilon}_\varphi\frac{\partial \tilde{\upsilon}_\varphi}{r\partial \varphi} + \rho\tilde{\upsilon}_z\frac{\partial \tilde{\upsilon}_\varphi}{\partial z} + \rho\sum_s \bar{\upsilon}_s\frac{\partial \tilde{\upsilon}_\varphi}{\partial x_s} = \eta\Delta\tilde{\upsilon}_\varphi + \tilde{F}_\varphi \\
\rho\frac{d\tilde{\upsilon}_z}{dt} &= \rho\frac{\partial \tilde{\upsilon}_z}{\partial t} + \rho\tilde{\upsilon}_r\frac{\partial \tilde{\upsilon}_z}{\partial r} + \rho\tilde{\upsilon}_\varphi\frac{\partial \tilde{\upsilon}_z}{r\partial \varphi} + \rho\tilde{\upsilon}_z\frac{\partial \tilde{\upsilon}_z}{\partial z} + \rho\sum_s \bar{\upsilon}_s\frac{\partial \tilde{\upsilon}_z}{\partial x_s} = \eta\Delta\tilde{\upsilon}_z + \tilde{F}_z
\end{aligned} \tag{4.2a}$$

where for a linear float motion, $\bar{\upsilon}_k$, we obtain similarly:

$$\rho\frac{\mathrm{d}\bar{\upsilon}_k}{\mathrm{d}t} = \rho\frac{\partial \bar{\upsilon}_k}{\partial t} + \rho\sum_n \bar{\upsilon}_n\frac{\partial \bar{\upsilon}_k}{\partial x_n} + \rho\tilde{\upsilon}_r\frac{\partial \bar{\upsilon}_k}{\partial r} + \rho\tilde{\upsilon}_\varphi\frac{\partial \bar{\upsilon}_k}{r\partial \varphi} + \rho\tilde{\upsilon}_z\frac{\partial \bar{\upsilon}_k}{\partial z} = \eta\Delta\bar{\upsilon}_k + \bar{F}_k \tag{4.2b}$$

while a simpler relation presents a system of transport equations related to the self-organized single vortex motion:

$$\begin{aligned}
\rho\frac{\mathrm{d}\tilde{\upsilon}_r}{\mathrm{d}t} &= \rho\frac{\partial \tilde{\upsilon}_r}{\partial t} + \rho\tilde{\upsilon}_r\frac{\partial \tilde{\upsilon}_r}{\partial r} + \rho\tilde{\upsilon}_\varphi\frac{\partial \tilde{\upsilon}_r}{r\partial \varphi} + \rho\tilde{\upsilon}_z\frac{\partial \tilde{\upsilon}_r}{\partial z} = \eta\Delta\tilde{\upsilon}_r + \tilde{F}_r \\
\rho\frac{\mathrm{d}\tilde{\upsilon}_\varphi}{\mathrm{d}t} &= \rho\frac{\partial \tilde{\upsilon}_\varphi}{\partial t} + \rho\tilde{\upsilon}_\varphi\frac{\partial \tilde{\upsilon}_\varphi}{\partial r} + \rho\tilde{\upsilon}_\varphi\frac{\partial \tilde{\upsilon}_\varphi}{r\partial \varphi} + \rho\tilde{\upsilon}_z\frac{\partial \tilde{\upsilon}_\varphi}{\partial z} = \eta\Delta\tilde{\upsilon}_\varphi + \tilde{F}_\varphi
\end{aligned}$$

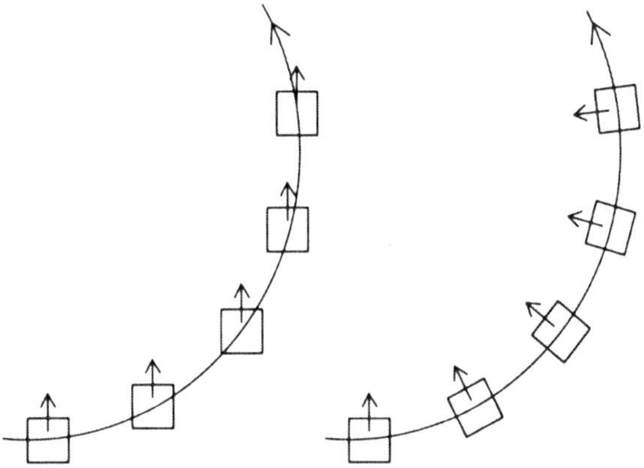

Fig. 4.1 Sketch of the transport trajectories with points having three degrees of freedom in space as relates to the vorticity (*left part*) and six degrees of freedom (*right part*); in a plane we observe only one degree of freedom on the *left side* and two degrees of freedom on the *right*

$$\rho\frac{\mathrm{d}\tilde{v}_z}{\mathrm{d}t} = \rho\frac{\partial \tilde{v}_z}{\partial t} + \rho\tilde{v}_r\frac{\partial \tilde{v}_z}{\partial r} + \rho\tilde{v}_\varphi\frac{\partial \tilde{v}_z}{r\partial\varphi} + \rho\tilde{v}_z\frac{\partial \tilde{v}_z}{\partial z} = \eta\Delta\tilde{v}_z + \tilde{F}_z \tag{4.2c}$$

The main difference between this approach and the classic vortex definition (4.1a, b, c) is demonstrated in Fig. 4.1; we indicate that the classical approach has only three degrees of freedom in space, instead of six ones as here.

Additionally, we may consider the combinations of the linear and vortex motions; we present some different combination of the basic motions and related float transports:

$$\begin{aligned}
&\rho\left(\frac{\partial}{\partial t} + \tilde{v}_r\frac{\partial}{\partial r} + \tilde{v}_\varphi\frac{\partial}{r\partial\varphi} + \tilde{v}_z\frac{\partial}{\partial z} + \sum_s \bar{v}_s\frac{\partial}{s}\right)\tilde{v}_{(r,\varphi,z)} = \eta\Delta\tilde{v}_{(r,\varphi,z)} + \tilde{F}_{(r,\varphi,z)}\\
&\rho\left(\frac{\partial}{\partial t} + \sum_s \bar{v}_s\frac{\partial}{s} + \tilde{v}_r\frac{\partial}{\partial r} + \tilde{v}_\varphi\frac{\partial}{r\partial\varphi} + \tilde{v}_z\frac{\partial}{}\right)\bar{v}_s = \eta\Delta\bar{v} + \bar{F}_s
\end{aligned} \tag{4.3a}$$

describing interrelation between vortex and linear motion and for vortex and rotation

$$\begin{aligned}
&\rho\left(\frac{\partial}{\partial t} + \tilde{v}_r\frac{\partial}{\partial r} + \tilde{v}_\varphi\frac{\partial}{r\partial\varphi} + \tilde{v}_z\frac{\partial}{\partial z} + v_r\frac{\partial}{\partial r} + v_\theta\frac{\partial}{r\partial\theta} + v_\psi\frac{\partial}{r\sin\theta\,\partial\psi}\right)\tilde{v}_{(r,\varphi,z)} = \eta\Delta\tilde{v}_{(r,\varphi,z)} + \tilde{F}_{(r,\varphi,z)}\\
&\rho\left(\frac{\partial}{\partial t} + v_r\frac{\partial}{\partial r} + v_\theta\frac{\partial}{r\partial\theta} + v_\psi\frac{\partial}{r\sin\theta\,\partial\psi} + \tilde{v}_r\frac{\partial}{\partial r} + \tilde{v}_\varphi\frac{\partial}{r\partial\varphi} + \tilde{v}_z\frac{\partial}{\partial z} + \right)v_{(r,\theta,\psi)} = \eta\Delta v_{(r,\theta,\psi)} + F_{(r,\theta,\psi)}
\end{aligned} \tag{4.3b}$$

Let us repeat that while dealing with two or more variable motion fields, e.g., v_k and $\bar{v}_k$, we must consider two or more transport relations: not only for v_k, but also for the other transport fields, e.g., $\bar{v}_k$.

Further we may include the changes of density; for example, we may consider the continuity relation for the two flows, e.g., transport term v_s, and float transport term $\bar{v}_s$:

$$\frac{d\rho}{dt} = \frac{\partial\rho}{\partial t} + \sum_s \frac{\partial(\rho\upsilon_s)}{\partial x_s} + \sum_s \frac{\partial(\rho\bar{\upsilon}_s)}{\partial x_s} = 0 \tag{4.4}$$

The transport field related to the vortex motion can be defined basing on the angular molecular moment field and internal spin motions and a certain Cosserat length:

$$\Omega_{(r,\varphi,z)} = l\left\{\left(\frac{\partial\omega_\varphi}{\partial z} - \frac{\partial\omega_z}{r\partial\varphi}\right), \left(\frac{\partial\omega_z}{\partial r} - \frac{\partial\omega_r}{\partial z}\right), \left(\frac{\partial\omega_r}{r\partial\varphi} - \frac{\partial\omega_\varphi}{\partial r}\right)\right\} \tag{4.5}$$

Hence we may define the angle related displacement velocity with the help of some effective arm, L_k:

$$\breve{\upsilon}_p = \varepsilon_{pks}\Omega_k L_s \tag{4.6}$$

The asymmetric molecular strain theory with a definition of rotational velocity transport forms the new basis for constructing a model of some disturbances in a non-laminar flow, like vortices.

The interaction molecular processes related to the vortex mechanism are studied with the help of a new definition of an asymmetric momentum flux.

In Chap. 3 we have discussed the motion equations for transport, float and variable reference systems; here we will present some simplified solutions for the vortex and related motions. We follow the Asymmetric Continuum Theory developed in our former papers and books (Teisseyre 2008a, b, 2009, 2011) and new attempts related to the Asymmetric Molecular Strain Theory (Teisseyre 2009, 2011). The first ones of the above-mentioned publications, related to seismology, present essential improvements in solid state continuum theory and interpretation of seismic data. Working in this direction we have already presented a new approach related to vorticity problems: instead of its former definition as a rotation of transport velocity, $\varsigma = \text{rot}\,\bar{\upsilon}$, we have proposed to consider the independent rotational motions related to the new definition of an angular moment with a variable arm and spin incorporated in the Asymmetric Molecular Strain Theory (Teisseyre 2010). Such an approach well describes the vortex transport phenomena based on this new theory.

Several papers based on relations between solids and fluids and on the theoretical approach to the non-laminar flows and vortices point out some regularities (e.g., Beirão da Veiga 2007; Fan et al. 2011; Farwig et al. 2009) revealed in the solutions for the non-laminar flows and vortices. However, we should mention that the present book, based on Asymmetric Continuum Theory in solids, introduces a quite new approach into the fluids mechanics.

Also, we can consider different transport structures, in particular, the ones including separate transport velocities for the quasi-linear transport motions, $\bar{\upsilon}$, and the vortex rotation transport, $\breve{\upsilon}$, related to the vortex structure. Thus, instead of the classical approach to vorticity as rotation of velocity transport we will rely on the linear transport and on rotation transport, $\tilde{\upsilon}_s = \{\upsilon_r, r\upsilon_\varphi, \upsilon_z\}$, related to a moment rate and presenting directly a vortex transport process (Teisseyre 2009). These vortex motions may influence a linear transport:

$$\frac{\mathrm{d}(\rho\upsilon_i)}{\mathrm{d}t} = \upsilon_i\frac{\partial\rho}{\partial t} + \rho\frac{\partial\upsilon_i}{\partial t} + \rho\sum_s \tilde{\upsilon}_s\frac{\partial\upsilon_i}{\partial x_s} + \sum_k \upsilon_k\frac{\partial\upsilon_i}{\partial x_k} = \eta\sum_k\frac{\partial^2\upsilon_i}{\partial x_k\partial x_k} + F_i \quad (4.7a)$$

and reversely the transport related to a vortex structure, $\tilde{\upsilon}$, may be influenced by the linear transport term, υ_k:

$$\frac{\mathrm{d}\left(\rho\tilde{\upsilon}_i\right)}{\mathrm{d}t} = \tilde{\upsilon}_i\frac{\partial\rho}{\partial t} + \rho\frac{\partial\tilde{\upsilon}_i}{\partial t} + \rho\sum_s \tilde{\upsilon}_s\frac{\partial\tilde{\upsilon}_i}{\partial x_s} + \sum_k \upsilon_k\frac{\partial\tilde{\upsilon}_i}{\partial x_k} = \eta\Delta\tilde{\upsilon}_i + F_i \quad (4.7b)$$

In this way we may describe some deformations of a vortex structure by linear transport, e.g., presenting an influence of wind on a vortex motion in the atmosphere. Further, following our approach applied to the antisymmetric theory in solids we may have, along with the symmetric molecular strains, also the antisymmetric molecular strains (strain rates) and related molecular stresses

$$\tilde{E}_{kl} = \tilde{E}_{(kl)} + \tilde{E}_{[kl]}\ ,\ \tilde{S}_{kl} = \tilde{S}_{(kl)} + \tilde{S}_{[kl]} \quad (4.8a)$$

For simplicity, we use the following constitutive relations:

$$\tilde{S}_{(kl)} = \tilde{\eta}\tilde{E}_{(kl)}\ ,\ \tilde{S}_{[kl]} = \tilde{\eta}\tilde{E}_{[kl]} \quad \text{at} \quad \frac{1}{3}\sum_s\tilde{S}_{ss} = -\frac{\partial p}{\partial t} = \frac{1}{3}k\sum_s\tilde{E}_{ss} \approx 0 \quad (4.8b)$$

where $\tilde{\eta}$ is the related viscosity constant, and the small pressure variations, $\frac{\delta_{kl}}{3}\sum_s\tilde{S}_{(ss)} \approx \frac{\partial p}{\partial t}$; these sound waves could be in our considerations neglected.

In so defined fluids, the point motions have six degrees of freedom: displacement velocity and rotation motion – spin. The independent molecular strains can be related to existing displacement velocities; let us assume the existence of some transport field, υ_l. Then we may write for the related molecular strains (strain time derivatives):

$$\tilde{E}_{kl} = \frac{1}{2}\left(\frac{\partial\upsilon_l}{\partial x_k} + \frac{\partial\upsilon_k}{\partial x_l}\right)\ ,\ \tilde{E}_{[kl]} = \frac{1}{2}\left[\frac{\partial\upsilon_l}{\partial x_k} - \frac{\partial\upsilon_k}{\partial x_l}\right] \quad (4.9)$$

These molecular strains, shear and rotation rates, can be treated as independent fields, while the related displacements help us only to derive the independent motion equations for these molecular strains. In so defined fluids, these point motions have nine degrees of freedom: displacement velocity, shear and rotation motions.

Next, we consider the momentum flux as related to the new asymmetric form. For the molecular symmetric stresses we can write, after Landau and Lifshitz (1959), new edition translated into Polish as Landau and Lifszyc (2009):

$$\tilde{S}_{(ij)} = 2\eta\left(\tilde{E}_{(ij)} - \frac{\delta_{ij}}{3}\sum_s \tilde{E}_{ss}\right) + \varepsilon\delta_{ij}\sum_s \tilde{E}_{ss}, \text{ or}$$
$$\tilde{S}_{(ij)} = \eta\left(\frac{\partial \upsilon_j}{\partial x_i} + \frac{\partial \upsilon_i}{\partial x_j} - \frac{2\delta_{ij}}{3}\sum_s \frac{\partial \upsilon_s}{\partial x_s}\right) + \varepsilon\delta_{ij}\sum_s \frac{\partial \upsilon_s}{\partial x_s} \tag{4.10a}$$

In our new approach we should add an influence of rotation molecular processes; therefore, we introduce the asymmetric fluid viscous stress tensor adding an influence of the antisymmetric molecular stresses (Teisseyre 2009):

$$\tilde{S}_{[ij]} = 2\eta\tilde{E}_{[ij]} = \eta\left(\frac{\partial \upsilon_j}{\partial x_i} - \frac{\partial \upsilon_i}{\partial x_j}\right) \tag{4.10b}$$

The total molecular stresses become given as:

$$\tilde{S}_{ij} = \tilde{S}_{(ij)} + \tilde{S}_{[ij]}; \tilde{S}_{ij} = 2\eta\left(\frac{\partial \upsilon_j}{\partial x_i} - \frac{\delta_{ij}}{3}\sum_s \frac{\partial \upsilon_s}{\partial x_s}\right) + \varepsilon\delta_{ij}\sum_s \frac{\partial \upsilon_s}{\partial x_s} \tag{4.10c}$$

Now, according to Landau and Lifshitz (1959), new edition translated into Polish as Landau and Lifszyc (2009), we can write for the modified asymmetric momentum flux:

$$T_{(kn)} = -\tilde{S}_{(kn)} + \rho\,\upsilon_k\,\upsilon_n, T_{[kn]} = -\tilde{S}_{[kn]}; \upsilon = \{\tilde{r}, r\tilde{\omega}, \tilde{z}\} \tag{4.11a}$$

Note here a special definition applied for a rotation transport field, $\{\tilde{r}, r\tilde{\omega}, \tilde{z}\}$, as slightly different from that following from a simple coordinate transformation (we put $r\tilde{\omega}$ instead of $\tilde{r}\tilde{\omega}$; this brings an important consequence to our considerations).

We write

$$T_{kn} = -2\eta\left(\frac{\partial \upsilon_n}{\partial x_k} - \frac{\delta_{kn}}{3}\sum_s \frac{\partial \upsilon_s}{\partial x_s}\right) - \varepsilon\delta_{kn}\sum_s \frac{\partial \upsilon_s}{\partial x_s} + \rho\upsilon_k\upsilon_n \tag{4.11b}$$

arriving at a general expression for the motion of fluids which reads as follows

$$\frac{\partial(\rho\upsilon_n)}{\partial t} + \sum_k \frac{\partial T_{kn}}{\partial x_k} = \rho\tilde{F}_n \tag{4.12a}$$

and we obtain:

$$\upsilon_n\frac{\partial \rho}{\partial t} + \rho\frac{\partial \upsilon_n}{\partial t} - 2\eta\sum_k \frac{\partial^2 \upsilon_n}{\partial x_k \partial x_k} + \left(\frac{2\eta}{3} - \varepsilon\right)\sum_s \frac{\partial^2 \upsilon_s}{\partial x_n \partial x_s} +$$
$$\upsilon_n\sum_k \upsilon_k\frac{\partial \rho}{\partial x_k} + \rho\upsilon_n\sum_k \frac{\partial \upsilon_k}{\partial x_k} + \rho\sum_k \upsilon_k\frac{\partial \upsilon_n}{\partial x_k} = \rho\tilde{F}_n \tag{4.12b}$$

For the mass conservation we have

$$\frac{d\rho}{dt} = \frac{\partial \rho}{\partial t} + \sum_s \frac{\partial(\rho \upsilon_s)}{\partial x_s} = 0 \tag{4.12c}$$

Further, let us consider some specific problems, namely the opposite-direction flows around a stationary vortex; we will consider this problem only in a simplified way. We assume that some simple model may describe the uniform transport related to two uniform opposite-direction flows; the equations for stationary molecular strains can be found using Eqs. (4.11a) and (4.12a) outside the rotation region, that is, at $\tilde{\upsilon} = 0$, and under a constant thermodynamical condition, $\frac{\partial \rho}{\partial t} = 0$:

$$\begin{gathered} -2\eta \sum_k \frac{\partial^2 \upsilon_n}{\partial x_k \partial x_k} + \left(\frac{2\eta}{3} - \varepsilon\right) \sum_s \frac{\partial^2 \upsilon_s}{\partial x_n \partial x_s} + \\ \upsilon_n \sum_k \upsilon_k \frac{\partial \rho}{\partial x_k} + \rho \upsilon_n \sum_k \frac{\partial \upsilon_k}{\partial x_k} + \rho \sum_k \upsilon_k \frac{\partial \upsilon_n}{\partial x_k} = \rho \tilde{F}_n \end{gathered} \tag{4.13a}$$

and for the mass conservation:

$$\sum_s \frac{\partial(\rho \upsilon_s)}{\partial x_s} = 0 \tag{4.13b}$$

For a single stationary vortex, we consider a mechanical model near a boundary of these two opposite-direction uniform flows; moreover, we may consider also a row of vortices with some decrease of the vortex radii from R_{-3} to R_3 along the x line at $z = 0$: $R \to R_{-3} \le R_{-2} \le R_{-1} \le R \ge R_1 \ge R_2 \ge R_3$.

At the boundary of the opposite-direction flows, the relative transport velocity may overpass the Reynolds critical number and we may expect the appearance of a vortex or vortices. Further on, we will consider a simple mechanical model of the vortex motions near a boundary of these two opposite-direction uniform flows. Outside the vortices we may have a curvilinear system of flow around and between the vortices.

First we will concentrate on a stationary vortex motion; a macroscopic transport field, υ, can be treated as an imposed external field, while the internal transport rotation will enter into the additional terms introduced to the system of equations. Thus, in motions with advanced vorticity dynamics we assume that transport can be related to both the displacement velocity and the vortex motions; we follow the approach applied to fragmentation and slip in the fracture processes (Teisseyre and Górski 2009a, b); in such approach the rotation transport may sufficiently describe the vortex motions (Teisseyre 2009).

For a large Reynolds number, the laminar motions become unstable and the dual transport process with displacement velocities and spin motions may generate the micro-vortices; a kind of dynamic vortex structure can be formed, simultaneously undergoing a displacement transport process. However, under some special conditions, an isolated vortex center can be formed; inside it, the displacement velocity transport might be negligible. However, to consider formally the transport related to such extreme conditions we should return to the basic problem of how to incorporate the spin motion into the system of basic relations.

Following the paper by Teisseyre (2009), the rotation transport υ_s can be given as radial velocity, $\tilde{r}$, angular velocity, $\tilde{\omega}$, and axial velocity, $\tilde{z}$:

$$\upsilon_k \rightarrow \{\tilde{r}, r\tilde{\omega}, \tilde{z}\} \tag{4.14}$$

We obtain for a rotation transport in the vortices along the z-axis:

$$\begin{aligned}
&\frac{\mathrm{d}}{\mathrm{d}t} = \frac{\partial}{\partial t} + \tilde{\upsilon}_k \frac{\partial}{\partial x_k} \rightarrow \frac{\mathrm{d}}{\mathrm{d}t} = \frac{\partial}{\partial t} + \tilde{r}\frac{\partial}{\partial r} + \tilde{\omega}\frac{\partial}{\partial \varphi} + \tilde{z}\frac{\partial}{\partial z} \rightarrow \\
&\frac{\mathrm{d}(\rho\tilde{r})}{\mathrm{d}t} = \tilde{r}\frac{\partial \rho}{\partial t} + \rho\frac{\partial \tilde{r}}{\partial t} + \rho\tilde{r}\frac{\partial \tilde{r}}{\partial r} + \rho\tilde{\omega}\frac{\partial \tilde{r}}{\partial \varphi} + \rho\tilde{z}\frac{\partial \tilde{r}}{\partial z} \\
&\frac{\mathrm{d}(\rho\tilde{\omega})}{\mathrm{d}t} = \tilde{\omega}\frac{\partial \rho}{\partial t} + \rho\frac{\partial \tilde{\omega}}{\partial t} + + \rho\tilde{r}\frac{\partial \tilde{\omega}}{\partial r} + \rho\tilde{\omega}\frac{\partial \tilde{\omega}}{\partial \varphi} + \rho\tilde{z}\frac{\partial \tilde{\omega}}{\partial z} \\
&\frac{\mathrm{d}(\rho\tilde{z})}{\mathrm{d}t} = \tilde{z}\frac{\partial \rho}{\partial t} + \rho\frac{\partial \tilde{z}}{\partial t} + + \rho\tilde{r}\frac{\partial \tilde{z}}{\partial r} + \rho\tilde{\omega}\frac{\partial \tilde{z}}{\partial \varphi} + \rho\tilde{z}\frac{\partial \tilde{z}}{\partial z}
\end{aligned} \tag{4.15}$$

At the vortex motion, from Eq. (4.11a, b) we obtain for $T_{kn} = \rho\tilde{\upsilon}_k\tilde{\upsilon}_n$, $(\tilde{\upsilon} = \{\tilde{r}, \tilde{\omega}, \tilde{z}\})$:

$$\frac{\partial(\rho\upsilon_n)}{\partial t} + \sum_k \frac{\partial T_{kn}}{\partial x_k} = \rho\tilde{F}_n, T_{kn} = \rho\tilde{\upsilon}_k\tilde{\upsilon}_n \tag{4.16}$$

that is

$$\begin{aligned}
\frac{\partial(\rho\tilde{r})}{\partial t} + \frac{\partial(\rho\tilde{r}^2)}{\partial r} + \frac{\partial(\rho\tilde{\omega}\tilde{r})}{\partial \varphi} + \frac{\partial(\rho\tilde{r}\tilde{z})}{\partial z} &= \rho\tilde{F}_n \\
\frac{\partial(\rho\tilde{\omega})}{\partial t} + \frac{\partial(\rho\tilde{r}\tilde{\omega})}{\partial r} + \frac{\partial(\rho\tilde{\omega}^2)}{\partial \varphi} + \frac{\partial(\rho\tilde{\omega}\tilde{z})}{\partial z} &= \rho\tilde{M}_n \\
\frac{\partial(\rho\tilde{z})}{\partial t} + \frac{\partial(\rho\tilde{r}\tilde{z})}{\partial r} + \frac{\partial(\rho\tilde{\omega}\tilde{z})}{\partial \varphi} + \frac{\partial(\rho\tilde{z}^2)}{\partial z} &= \rho\tilde{N}_n
\end{aligned} \tag{4.17a}$$

where we have defined: $\tilde{F}_n = \{\tilde{F}, \tilde{M}, \tilde{N}\}$.

For the mass conservation we have

$$\frac{d\rho}{dt} = \frac{\partial \rho}{\partial t} + \tilde{r}\frac{\partial \rho}{\partial r} + \tilde{\omega}\frac{\partial \rho}{\partial \varphi} + \tilde{z}\frac{\partial \rho}{\partial z} = 0 \tag{4.17b}$$

These relations present the dynamic situation under a given thermodynamical condition. At constant thermodynamical condition, $\frac{\partial \rho}{\partial t} = 0$, we obtain for the stationary relations:

$$\begin{aligned}
\frac{\partial(\rho\tilde{r}^2)}{\partial r} + \frac{\partial(\rho\tilde{\omega}\tilde{r})}{\partial\varphi} + \frac{\partial(\rho\tilde{r}\tilde{z})}{\partial z} &= \tilde{F} \\
\frac{\partial(\rho\tilde{r}\tilde{\omega})}{\partial r} + \frac{\partial(\rho\tilde{\omega}^2)}{\partial\varphi} + \frac{\partial(\rho\tilde{\omega}\tilde{z})}{\partial z} &= \tilde{M} \\
\frac{\partial(\rho\tilde{r}\tilde{z})}{\partial r} + \frac{\partial(\rho\tilde{\omega}\tilde{z})}{\partial\varphi} + \frac{\partial(\rho\tilde{z}^2)}{\partial z} &= \tilde{N}
\end{aligned} \tag{4.18a}$$

and for the mass conservation condition related to the velocity field we have:

$$\tilde{r}\frac{\partial\rho}{\partial r} + \tilde{\omega}\frac{\partial\rho}{\partial\varphi} + \tilde{z}\frac{\partial\rho}{\partial z} = 0 \tag{4.18b}$$

This complex stationary system may be solved for two cases; first with constant density, $\rho = \rho_0$ (e.g., for liquids), and then with a variable density inside the vortices, $\rho(r,\varphi,z)$; still, this complex system will remain quite complicated. First, let us consider the stationary vortex relations at the constant density and confine our consideration to a single vortex; taking into account (4.18a), the related stationary equations at constant thermodynamical condition with $\rho = \text{const}$ become:

$$\begin{aligned}
\frac{\partial(\tilde{r}^2)}{\partial r} + \frac{\partial(\tilde{\omega}\tilde{r})}{\partial\varphi} + \frac{\partial(\tilde{r}\tilde{z})}{\partial z} &= \tilde{F} \\
\frac{\partial(\tilde{r}\tilde{\omega})}{\partial r} + \frac{\partial(\tilde{\omega}^2)}{\partial\varphi} + \frac{\partial(\tilde{\omega}\tilde{z})}{\partial z} &= \tilde{M} \\
\frac{\partial(\tilde{r}\tilde{z})}{\partial r} + \frac{\partial(\tilde{\omega}\tilde{z})}{\partial\varphi} + \frac{\partial(\tilde{z}^2)}{\partial z} &= \tilde{N}
\end{aligned} \tag{4.19}$$

where we have added in explicit form the applied constant force F and moment M.

Note that such flows can occur for the Reynolds numbers above the critical value.

For a constant density we can demand:

$$\begin{aligned}
&\tilde{r} = \tilde{r}_0\exp(\gamma r)\exp(\varepsilon\varphi)\exp(\alpha z), \tilde{\omega} = \tilde{\omega}_0\exp(\gamma r)\exp(\varepsilon\varphi)\exp(\alpha z), \\
&\tilde{z} = \tilde{z}_0\exp(\gamma r)\exp(\varepsilon\varphi)\exp(\alpha z), \rho = \rho_0
\end{aligned} \tag{4.20}$$

and we obtain the conditions:

$$\begin{aligned}
&2\tilde{r}\frac{\partial\tilde{r}}{\partial r} + \tilde{r}\frac{\partial\tilde{\omega}}{\partial\varphi} + \tilde{r}\frac{\partial\tilde{z}}{\partial z} + \tilde{\omega}\frac{\partial\tilde{r}}{\partial\varphi} + \tilde{z}\frac{\partial\tilde{r}}{\partial z} = F \rightarrow \\
&2(\gamma\tilde{r}_0^2 + \varepsilon\tilde{r}_0\tilde{\omega}_0 + \alpha\tilde{z}_0\tilde{r}_0)\exp(2\gamma r)\exp(2\varepsilon\varphi)\exp(2\alpha z) = \tilde{F}
\end{aligned} \tag{4.21a}$$

$$\begin{aligned}
&2\tilde{\omega}\frac{\partial\tilde{\omega}}{\partial\varphi} + \tilde{\omega}\frac{\partial\tilde{r}}{\partial r} + \tilde{r}\frac{\partial\tilde{\omega}}{\partial r} + \tilde{z}\frac{\partial\tilde{\omega}}{\partial z} + \tilde{\omega}\frac{\partial\tilde{z}}{\partial z} = M \rightarrow \\
&2(\varepsilon\tilde{\omega}_0^2 + \gamma\tilde{\omega}_0\tilde{r}_0 + \alpha\tilde{\omega}_0\tilde{z}_0)\exp(2\gamma r)\exp(2\varepsilon\varphi)\exp(2\alpha z) = \tilde{M}
\end{aligned} \tag{4.21b}$$

and

$$2\tilde{z}\frac{\partial \tilde{z}}{\partial z} + \tilde{z}\frac{\partial \tilde{r}}{\partial r} + \tilde{r}\frac{\partial \tilde{z}}{\partial r} + \tilde{\omega}\frac{\partial \tilde{z}}{\partial \varphi} + \tilde{z}\frac{\partial \tilde{\omega}}{\partial \varphi} = \tilde{N} \rightarrow$$
$$2(\alpha\tilde{z}_0^2 + \gamma\tilde{r}_0\tilde{z}_0 + \varepsilon\tilde{\omega}_0\tilde{z}_0)\exp(2\gamma r)\exp(2\varepsilon\varphi)\exp(2\alpha z) = \tilde{N} \tag{4.21c}$$

The obtained conditions state that the coefficients in the exponents should be the same for all fields.

Further we obtain:

$$\begin{aligned}
&\left(\gamma\tilde{r}_0^2 + \varepsilon\tilde{r}_0\tilde{\omega}_0 + \alpha\tilde{z}_0\tilde{r}_0\right)\exp[2(\gamma r + \varepsilon\varphi + \alpha z)] = \frac{1}{2}\tilde{F}\\
&\rightarrow \gamma\tilde{r}_0^2 + \varepsilon\tilde{r}_0\tilde{\omega}_0 + \alpha\tilde{z}_0\tilde{r}_0 = \frac{1}{2}F \text{ and } \gamma r + \varepsilon\varphi + \alpha z = 0\\
&(\varepsilon\tilde{\omega}_0^2 + \gamma\tilde{\omega}_0\tilde{r}_0 + \alpha\tilde{\omega}_0\tilde{z}_0)\exp[2(\gamma r + \varepsilon\varphi + \alpha z)] = \frac{1}{2}\tilde{M}\\
&\rightarrow \varepsilon\tilde{\omega}_0^2 + \gamma r_0\tilde{\omega}_0 + \alpha\tilde{\omega}_0\tilde{z} = \frac{1}{2}M \text{ and } \gamma r + \varepsilon\varphi + \alpha z = 0\\
&\left(\alpha\tilde{z}_0^2 + \gamma\tilde{r}_0\tilde{z}_0 + \varepsilon\tilde{\omega}_0\tilde{z}_0\right)\exp[2(\gamma r + \varepsilon\varphi + \alpha z)] = \frac{1}{2}\tilde{N}\\
&\rightarrow \alpha\tilde{z}_0^2 + \gamma\tilde{r}_0\tilde{z}_0 + \varepsilon\tilde{\omega}_0\tilde{z}_0 = \frac{1}{2}g \text{ and } \gamma r + \varepsilon\varphi + \alpha z = 0
\end{aligned} \tag{4.22}$$

The last equation becomes extremely important, being a final element of our solution; it means that points in a vortex motion merge due to the appearance of special molecular bonds:

$$\gamma r + \varepsilon\varphi + \alpha z = 0 \tag{4.23}$$

We add the boundary condition at $z = 0$ (cf., Eq. 4.14):

$$\begin{aligned}
&\upsilon_k = \{\tilde{r}, r\tilde{\omega}, \tilde{z}\} \rightarrow \{\tilde{R}, R\tilde{\omega}, \tilde{z}\}\\
&R = R_0 - r \text{ and } \upsilon_\varphi = R\tilde{\omega}
\end{aligned} \tag{4.24}$$

meaning that the radius of vortex, $R = R_0 - r$, should decrease when r is increasing, while the $\tilde{z}$ field must decrease along the z axis; such conditions permit to fulfil Eq. (4.23). Equations (4.23) and (4.22) with (4.24) present together the required solution.

Next we may consider the stationary vortex relations at the variable density; for this problem we should solve the stationary case with a variable density and constant thermodynamical condition. From Eq. (4.18a) we have for the stationary conditions, similarly to Eq. (4.21a):

$$\begin{aligned}
&\tilde{r}^2\frac{\partial\rho}{\partial r}+\tilde{\omega}\tilde{r}\frac{\partial\rho}{\partial\varphi}+\tilde{r}\tilde{z}\frac{\partial\rho}{\partial z}+\rho\left(2\tilde{r}\frac{\partial\tilde{r}}{\partial r}+\tilde{r}\frac{\partial\tilde{\omega}}{\partial\varphi}+\tilde{\omega}\frac{\partial\tilde{r}}{\partial\varphi}+\tilde{r}\frac{\partial\tilde{z}}{\partial z}+\tilde{z}\frac{\partial\tilde{r}}{\partial z}\right)=\rho\tilde{F}\\
&\tilde{r}\tilde{\omega}\frac{\partial\rho}{\partial r}+\tilde{\omega}^2\frac{\partial\rho}{\partial\varphi}+\tilde{\omega}\tilde{z}\frac{\partial\rho}{\partial z}+\rho\left(2\tilde{\omega}\frac{\partial\tilde{\omega}}{\partial\varphi}+\tilde{\omega}\frac{\partial\tilde{r}}{\partial r}+\tilde{r}\frac{\partial\tilde{\omega}}{\partial r}+\tilde{z}\frac{\partial\tilde{\omega}}{\partial z}+\tilde{\omega}\frac{\partial\tilde{z}}{\partial z}\right)=\rho\tilde{M}\\
&\tilde{r}\tilde{z}\frac{\partial\rho}{\partial r}+\tilde{\omega}\tilde{z}\frac{\partial\rho}{\partial\varphi}+\tilde{z}^2\frac{\partial\rho}{\partial z}+\rho\left(2\tilde{z}\frac{\partial\tilde{z}}{\partial z}+\tilde{z}\frac{\partial\tilde{r}}{\partial r}+\tilde{r}\frac{\partial\tilde{z}}{\partial r}+\tilde{\omega}\frac{\partial\tilde{z}}{\partial\varphi}+\tilde{z}\frac{\partial\tilde{\omega}}{\partial\varphi}\right)=\rho\tilde{N}
\end{aligned}\tag{4.25}$$

and for the mass conservation we have:

$$\tilde{r}\frac{\partial\rho}{\partial r}+\tilde{\omega}\frac{\partial\rho}{\partial\varphi}+\tilde{z}\frac{\partial\rho}{\partial z}=0\tag{4.26}$$

Using this assumption (Eq. 4.26) for $\tilde{r},\tilde{\omega},\tilde{z}$, we can consider the following types of relations:

$$\begin{aligned}
&\tilde{r}=\tilde{r}_0\exp(\gamma r)\exp(\varepsilon\varphi)\exp(\alpha z),\quad \tilde{\omega}=\tilde{\omega}_0\exp(\gamma r)\exp(\varepsilon\varphi)\exp(\alpha z),\\
&\tilde{z}=\tilde{z}_0\exp(\gamma r)\exp(\varepsilon\varphi)\exp(\alpha z),\quad \rho=\rho_0\exp[\vartheta^r r+\vartheta^\varphi\varphi+\vartheta^z z]
\end{aligned}\tag{4.27}$$

From Eq. (4.27) we obtain, with the help of Eqs. (4.22), (4.23) and (4.25), the simple conditions:

$$\begin{aligned}
&\vartheta^r\tilde{r}_0^2+\vartheta^\varphi\tilde{\omega}_0\tilde{r}_0+\vartheta^z\tilde{r}_0\tilde{z}_0+2\left(\gamma\tilde{r}_0^2+\varepsilon\tilde{r}_0\tilde{\omega}_0+\alpha\tilde{z}_0\tilde{r}_0\right)=\tilde{F}\\
&\vartheta^r\tilde{r}_0\tilde{\omega}_0+\vartheta^\varphi\tilde{\omega}_0^2+\vartheta^z\tilde{\omega}_0\tilde{z}_0+2\left(\varepsilon\tilde{\omega}_0^2+\gamma\tilde{r}_0\tilde{\omega}_0+\alpha\tilde{\omega}_0\tilde{z}_0\right)=\tilde{M}\\
&\vartheta^r\tilde{r}_0\tilde{z}_0+\vartheta^\varphi\tilde{\omega}_0\tilde{z}_0+\vartheta^z\tilde{z}_0^2+2\left(\alpha\tilde{z}_0^2+\gamma\tilde{r}_0\tilde{z}_0+\varepsilon\tilde{\omega}_0\tilde{z}_0\right)=\tilde{N}
\end{aligned}\tag{4.28}$$

Due to the density conditions (4.26) we should supplement these relations by the relation:

$$\vartheta^r\tilde{r}_0+\vartheta^z\tilde{\omega}_0+\vartheta^z\tilde{z}_0=0\tag{4.29a}$$

However, we obtain

$$\begin{aligned}
&\tilde{r}\frac{\partial\rho}{\partial r}+\tilde{\omega}\frac{\partial\rho}{\partial\varphi}+\tilde{z}\frac{\partial\rho}{\partial z}\equiv\left(\tilde{r}\frac{\partial\rho}{\partial r}+\tilde{\omega}\frac{\partial\rho}{\partial\varphi}+\tilde{z}\frac{\partial\rho}{\partial z}\right)\rho_0\exp[\vartheta^r r+\vartheta^\varphi\varphi+\vartheta^z z]=0\\
&\quad\rightarrow\ \vartheta^r r+\vartheta^\varphi\varphi+\vartheta^z z=0
\end{aligned}\tag{4.29b}$$

Comparing the condition $\vartheta^r r+\vartheta^\varphi\varphi+\vartheta^z z=0$ with that given in Eq. (4.23) we put:

$$\vartheta^r=\gamma,\vartheta^\varphi=\varepsilon,\vartheta^z=\alpha\tag{4.29c}$$

Now, we consider the dynamic vortex relations; from Eqs. (4.17a, b) and (4.26–4.27) we have:

$$\begin{aligned}
&\tilde{r}\left(\frac{\partial\rho}{\partial t}+\tilde{r}\frac{\partial\rho}{\partial r}+\tilde{\omega}\frac{\partial\rho}{\partial\varphi}+\tilde{z}\frac{\partial\rho}{\partial z}\right)+\rho\left(\frac{\partial\tilde{r}}{\partial t}+2\tilde{r}\frac{\partial\tilde{r}}{\partial r}+\tilde{r}\frac{\partial\tilde{\omega}}{\partial\varphi}+\tilde{\omega}\frac{\partial\tilde{r}}{\partial\varphi}+\tilde{r}\frac{\partial\tilde{z}}{\partial z}+\tilde{z}\frac{\partial\tilde{r}}{\partial z}\right)=\rho\tilde{F}\\
&\tilde{\omega}\left(\frac{\partial\rho}{\partial t}+\tilde{r}\frac{\partial\rho}{\partial r}+\tilde{\omega}\frac{\partial\rho}{\partial\varphi}+\tilde{z}\frac{\partial\rho}{\partial z}\right)+\rho\left(\frac{\partial\tilde{\omega}}{\partial t}+2\tilde{\omega}\frac{\partial\tilde{\omega}}{\partial\varphi}+\tilde{\omega}\frac{\partial\tilde{r}}{\partial r}+\tilde{r}\frac{\partial\tilde{\omega}}{\partial r}+\tilde{z}\frac{\partial\tilde{\omega}}{\partial z}+\tilde{\omega}\frac{\partial\tilde{z}}{\partial z}\right)=\rho\tilde{M}\\
&\tilde{z}\left(\frac{\partial\rho}{\partial t}+\tilde{r}\frac{\partial\rho}{\partial r}+\tilde{\omega}\frac{\partial\rho}{\partial\varphi}+\tilde{z}\frac{\partial\rho}{\partial z}\right)+\rho\left(\frac{\partial\tilde{z}}{\partial t}+2\tilde{z}\frac{\partial\tilde{z}}{\partial z}+\tilde{z}\frac{\partial\tilde{r}}{\partial r}+\tilde{r}\frac{\partial\tilde{z}}{\partial r}+\tilde{\omega}\frac{\partial\tilde{z}}{\partial\varphi}+\tilde{z}\frac{\partial\tilde{\omega}}{\partial\varphi}\right)=\rho\tilde{N}
\end{aligned}\tag{4.30a}$$

and for the mass conservation:

$$\frac{d\rho}{dt}=\frac{\partial\rho}{\partial t}+\tilde{r}\frac{\partial\rho}{\partial r}+\tilde{\omega}\frac{\partial\rho}{\partial\varphi}+\tilde{z}\frac{\partial\rho}{\partial z}=0\tag{4.30b}$$

From these equations we get:

$$\begin{aligned}
&\frac{\partial\tilde{r}}{\partial t}+2\tilde{r}\frac{\partial\tilde{r}}{\partial r}+\tilde{r}\frac{\partial\tilde{\omega}}{\partial\varphi}+\tilde{\omega}\frac{\partial\tilde{r}}{\partial\varphi}+\tilde{r}\frac{\partial\tilde{z}}{\partial z}+\tilde{z}\frac{\partial\tilde{r}}{\partial z}=\tilde{F}\\
&\frac{\partial\tilde{\omega}}{\partial t}+2\tilde{\omega}\frac{\partial\tilde{\omega}}{\partial\varphi}+\tilde{\omega}\frac{\partial\tilde{r}}{\partial r}+\tilde{r}\frac{\partial\tilde{\omega}}{\partial r}+\tilde{z}\frac{\partial\tilde{\omega}}{\partial z}+\tilde{\omega}\frac{\partial\tilde{z}}{\partial z}=\tilde{M}\\
&\frac{\partial\tilde{z}}{\partial t}+2\tilde{z}\frac{\partial\tilde{z}}{\partial z}+\tilde{z}\frac{\partial\tilde{r}}{\partial r}+\tilde{r}\frac{\partial\tilde{z}}{\partial r}+\tilde{\omega}\frac{\partial\tilde{z}}{\partial\varphi}+\tilde{z}\frac{\partial\tilde{\omega}}{\partial\varphi}=\tilde{N}
\end{aligned}\tag{4.31a}$$

and according to (4.30b):

$$\frac{\partial\rho}{\partial t}+\tilde{r}\frac{\partial\rho}{\partial r}+\tilde{\omega}\frac{\partial\rho}{\partial\varphi}+\tilde{z}\frac{\partial\rho}{\partial z}=0\tag{4.31b}$$

We can put, similarly to Eq. (4.28):

$$\begin{aligned}
&\tilde{r}=\tilde{r}_0\exp[\theta t+\gamma r+\varepsilon\varphi+\alpha z],\tilde{\omega}=\tilde{\omega}_0\exp[\theta t+\gamma r+\varepsilon\varphi+\alpha z],\\
&\tilde{z}=\tilde{z}_0\exp[\theta t+\gamma r+\varepsilon\varphi+\alpha z],\rho=\rho_0\exp[\theta t+\gamma r+\varepsilon\varphi+\alpha z]
\end{aligned}\tag{4.32a}$$

From (4.31b), at variable density, we arrive at the unique relation joining the points by the special molecular bonds in a vortex motion:

$$\theta\rho+\gamma\tilde{r}+\varepsilon\tilde{\varphi}+\alpha\tilde{z}=0\;\rightarrow\;\theta\rho_0+\gamma\tilde{r}_0+\varepsilon\tilde{\varphi}_0+\alpha\tilde{z}_0=0\tag{4.32b}$$

where a parameter θ can depend directly on the thermodynamical conditions.

Instead of (4.20) we write with (4.24)

$$\begin{aligned}
&\tilde{R}=\tilde{R}_0\exp[\theta t-\gamma R+\varepsilon\varphi+\alpha z],\tilde{\omega}=\tilde{\omega}_0\exp[\theta t-\gamma R+\varepsilon\varphi+\alpha z],\\
&\tilde{z}=\tilde{z}_0\exp[\theta t-\gamma R+\varepsilon\varphi+\alpha z],\rho=\rho_0\exp[\theta t-\gamma R+\varepsilon\varphi+\alpha z]
\end{aligned}\tag{4.33a}$$

And again we should add the boundary condition at $z = 0$ (cf., Eq. 4.14):

$$r = 0 \text{ , or } R = R_0 \text{ and } v_\varphi = R_0\tilde{\omega}_0 \tag{4.33b}$$

meaning that the radius of vortex, $R = R_0 - r$, should decrease when r is increasing, similarly as $\tilde{\omega}$ for $\alpha > 0$, while the $\tilde{z}$ field must decrease along the z axis with $\gamma < 0$ and $\varepsilon \approx 0$; such conditions lead to

$$\theta\rho_0 + \gamma\tilde{r}_0 + \alpha\tilde{z}_0 \approx 0 \rightarrow \theta\rho_0 + \gamma\tilde{R}_0 + \alpha\tilde{z}_0 \approx 0 \rightarrow \tilde{R} = \tilde{R}_0\exp[\theta t - \gamma R + \varepsilon\varphi + \alpha z] \tag{4.34a}$$

related to a modification of the basic formula (4.31a) at $R = R_0 - r$ and $v_\varphi = R\tilde{\omega}$:

$$\begin{aligned}
&\frac{\partial \tilde{R}}{\partial t} + 2\tilde{R}\frac{\partial \tilde{R}}{\partial R} + \tilde{R}\frac{\partial \tilde{\omega}}{\partial \varphi} + \tilde{\omega}\frac{\partial \tilde{R}}{\partial \varphi} + \tilde{R}\frac{\partial \tilde{z}}{\partial z} + \tilde{z}\frac{\partial \tilde{R}}{\partial z} = \tilde{F}\\
&\frac{\partial \tilde{\omega}}{\partial t} + 2\tilde{\omega}\frac{\partial \tilde{\omega}}{\partial \varphi} + \tilde{\omega}\frac{\partial \tilde{R}}{\partial R} + \tilde{R}\frac{\partial \tilde{\omega}}{\partial R} + \tilde{z}\frac{\partial \tilde{\omega}}{\partial z} + \tilde{\omega}\frac{\partial \tilde{z}}{\partial z} = \tilde{M}\\
&\frac{\partial \tilde{z}}{\partial t} + 2\tilde{z}\frac{\partial \tilde{z}}{\partial z} + \tilde{z}\frac{\partial \tilde{R}}{\partial R} + \tilde{R}\frac{\partial \tilde{z}}{\partial R} + \tilde{\omega}\frac{\partial \tilde{z}}{\partial \varphi} + \tilde{z}\frac{\partial \tilde{\omega}}{\partial \varphi} = \tilde{N}
\end{aligned} \tag{4.34b}$$

resulting in the conditions permitting to fulfil Eq. (4.32b) at $t = 0$:

$$\begin{aligned}
&\theta\rho + \gamma\tilde{r} + \varepsilon\tilde{\varphi} + \alpha\tilde{z} = 0 \rightarrow \theta\rho - \gamma\tilde{R} + \varepsilon\tilde{\varphi} + \alpha\tilde{z} = 0\\
\text{at } \varepsilon \approx 0: \; &\theta\rho - \gamma\tilde{R} + \alpha\tilde{z} = 0 \rightarrow \theta\rho_0 - \gamma\tilde{R}_0 + \alpha\tilde{z}_0 = 0
\end{aligned} \tag{4.34c}$$

In this way we approach a simple turbulence model; we may construct a very simple turbulence model assuming that the two uniform opposite-directed flows have very different temperatures and reinforce a vortex motion by the gas packets with a variable temperature and thus also density. Thus, the related boundary and thermodynamical conditions will have a primary influence on the density and rotation motion. This rotation motion could be directly influenced by the linear or curvilinear Navier-Stockes transport at the vortex boundary starting from the value $R_0\tilde{\omega} \rightarrow R_0\tilde{\omega}_0\exp[-\gamma R_0]$ (cf., 4.33a; $\varepsilon \approx 0$). Some reinforcement of these motions may be due to the variable thermodynamical conditions touching directly the rotational components.

Now, we present a model of a single turbulence with the opposite-direction stream flows and a vortex formed when the relative velocity overpasses a laminar range; some disturbances appear due to a fluid viscosity. A vortex with a radius R (on the plane $z = 0$) can appear at the point $x = 0, y = 0$.

Let us assume that the disturbances due to these opposite-direction flows on the plane $z = 0$ extend in the y direction only inside the interval ($-L$, L) and outside this interval we have again the uniform flow in the x–direction only (Fig. 4.2).

First let us assume an incompressible fluid; its flow on plane $z = 0$ and nearby the boundary $y = 0$ can be described by some disturbances due to a fluid viscosity.

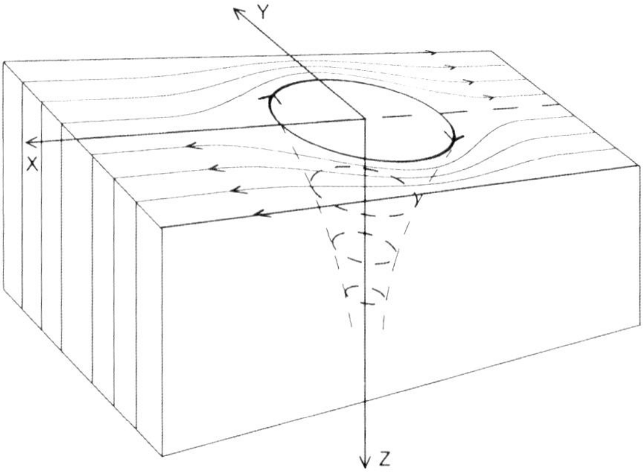

Fig. 4.2 Model of two opposite-direction flows and a vortex

We can consider the x and y components of this motion on plane $z = 0$ in the interval $Cx = (0 - 2\pi)$ (or $Cx = (0 - 2\pi), (2\pi - 4\pi), \ldots$) for a row of vortices, (Fig. 4.3) as follows:

$$\begin{aligned} \upsilon_x(x, y, z) &= \upsilon^0 \mathrm{sgn}(y) \exp(-A|y|) \exp(-Bz) \sqrt{\cos^2(Cx)} \\ \upsilon_y(x, y, z) &= \upsilon^0 \mathrm{sgn}(y) \exp(-A|y|) \exp(-Bz) \sqrt{\sin^2(Cx)} \\ \upsilon_z(y, z) &= \upsilon^0 \mathrm{sgn}(y) \exp(-A|y|) \exp(-Bz) \end{aligned} \tag{4.35}$$

For a compressible gas in the same interval, $Cx = (0 - 2\pi)$, we obtain:

$$\begin{aligned} \rho\upsilon_x(x, y, z) &= \rho^0 \upsilon^0 \mathrm{sgn}(y) \exp(-A|y|) \exp(-Bz) \sqrt{\cos^2(Cx)} \\ \rho\upsilon_y(x, y, z) &= \rho^0 \upsilon^0 \mathrm{sgn}(y) \exp(-A|y|) \exp(-Bz) \sqrt{\sin^2(Cx)} \\ \rho\upsilon_z(y, z) &= \rho^0 \upsilon^0 \mathrm{sgn}(y) \exp(-A|y|) \exp(-Bz) \end{aligned} \tag{4.36}$$

However, depending on the external condition we may correct the above relations putting the following expressions for υ_x , or $\rho\upsilon_x$:

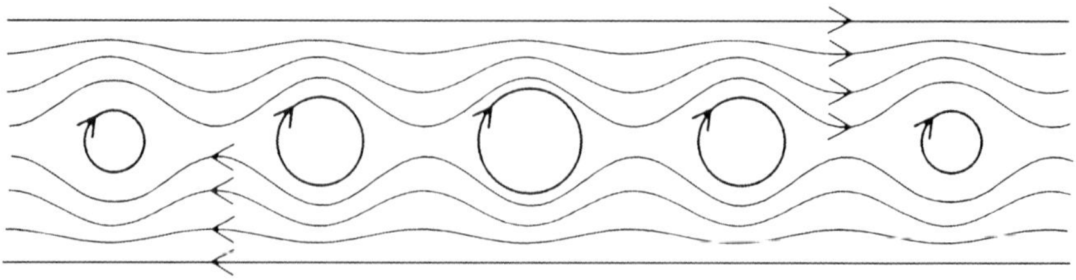

Fig. 4.3 Model of two opposite-direction flows and a row of vortices

$$v_x(x, y, z) = v^0 \mathrm{sgn}(y) \exp(-A|y|) \exp(-Bz) \sqrt{\cos^2(Cx) + D \sin^2(Cx)},$$
$$D = \{0, 1\} \tag{4.37a}$$

or

$$\rho v_x(x, y, z) = \rho_0 v^0 \mathrm{sgn}(y) \exp(-A|y|) \exp(-Bz) \sqrt{\cos^2(Cx) + D \sin^2(Cx)},$$
$$D = \{0, 1\} \tag{4.37b}$$

where due to the external condition the constant D may be even equal to 1; in this case we will have a constant mass flow in the x-direction:

$$v_x(x, y, z) = v^0 \mathrm{sgn}(y) \exp(-A|y|) \exp(-Bz) \sqrt{\cos^2(Cx) + D \sin^2(Cx)}, D = \{0, 1\} \tag{4.38a}$$

or

$$\rho v_x(x, y, z) = \rho_0 v^0 \mathrm{sgn}(y) \exp(-A|y|) \exp(-Bz) \sqrt{\cos^2(Cx) + D \sin^2(Cx)},$$
$$D = \{0, 1\} \tag{4.38b}$$

For simplicity we have considered the two opposite-direction flows and vortex motions almost separately. Now let us present some additional remarks how to join this approach with the local Reynolds numbers governing vortex formation.

We usually define the Reynolds number, Re, as the ratio of inertial to viscous forces; however, for our aim we may refer it to a ratio of the acceleration components. We propose to relate the Reynolds number to a ratio of an acceleration of a horizontal flow and the vertical gravity acceleration:

$$Re = Re^0 \frac{\dot{v}}{g} \tag{4.39a}$$

where Re^0 is the appropriate constant and for $Re > Re^0$ we will have a non-laminar flow, so we may expect the vortex processes.

On the other hand, when considering two opposite horizontal flows, the Reynolds numbers might be related to the angle, ψ, between actual horizontal flow direction in a plane $z = 0$ and the x-direction:

$$Re = Re^0(\psi - \psi_0) \tag{4.39b}$$

where according to this definition we may expect the vortex processes to occur for $\psi > \psi_0$.

To these considerations we should include also a material compressibility β; for a volume V_0 we write:

$$\beta = -\frac{1}{V_0}\frac{\partial V_0}{\partial p} \tag{4.40}$$

For an uncompressible material, $\beta = 0$, we put $\psi_0 = 0$ and under the conditions (4.39) we may expect an appearance of vortex motion for any angle $\psi > 0$.

For a compressible material we should add to the relations (4.39a, b) an additional factor related to compressibility,$\exp(-|\sigma\beta|)$, where the constant σ must be positive, $\sigma > 0$:

$$Re = Re^0 \frac{\dot{\upsilon}}{g} \exp(-|\sigma\beta|) = Re^0(\psi - \psi_0)\exp(-|\sigma\beta|) \tag{4.41}$$

and we may expect a decrease of the critical Reynolds number Re.

The presented approach permits to correct the formerly obtained results; first of all we should take into account that part of mass transport along the line x $(z = 0)$ may become directed to move downward in the z-direction forming a vortex motion:

$$\upsilon_k = \{\tilde{r}, r\tilde{\omega}, \tilde{z}\} \rightarrow \{\tilde{R}, R\tilde{\omega}, \tilde{z}\} \tag{4.42}$$

where $R = R_0 - r$ and $\upsilon_\varphi = R\tilde{\omega}$.

Further, for the accelerations $\dot{\upsilon} > g$, $(\dot{\upsilon} = g + \Delta\dot{\upsilon})$, a part of the acceleration ratio $\frac{\Delta\dot{\upsilon}}{g}$ might be converted into the vertical vortex motion, $\upsilon_k \rightarrow \{\tilde{r}, r\tilde{\omega}, \tilde{z}\}$ (cf., Eq. 4.24); due to this additional acceleration ratio, $\Delta\dot{\upsilon}/g$, some perturbances start and form the boundary condition for a vortex motion, $r = 0$, and $R = R_0$. This starting condition is related to some changes in the horizontal flow which become deviated from the x-direction up to the angle $\Delta\psi = \psi - \psi_0$; therefore, this horizontal mass flow will decrease from $\rho\upsilon_x(x_0, 0, 0) = \rho\upsilon^0 \mathrm{sgn}(y)$ up to the following value:

$$\begin{aligned} &\rho_0\upsilon_x(x_0, 0, 0) = \rho_0\upsilon^0 \mathrm{sgn}(y) \rightarrow \rho_0\upsilon_x(x_0, 0, 0) = \rho\upsilon^0 \mathrm{sgn}(y)\cos\Delta\psi, \\ &\sqrt{\cos^2(Cx_0)} = \cos\Delta\psi \end{aligned} \tag{4.43a}$$

or for a non-compressible material:

$$\begin{aligned} &\upsilon_x(x_0, 0, 0) = \upsilon^0 \mathrm{sgn}(y) \rightarrow \upsilon_x(x_0, 0, 0) = \upsilon^0 \mathrm{sgn}(y)\cos\Delta\psi, \\ &\sqrt{\cos^2(Cx_0)} = \cos\Delta\psi \end{aligned} \tag{4.43b}$$

In this way we can explain a direct supply of a vortex motion by the horizontal non-laminar flow.

Concluding our considerations we note that the main elements of the Asymmetric Continuum Theory for fluids are common both for solids and fluids. We believe that this Theory permits to avoid the main problems of the Classic Theory and, on the other hand, will remain to be based on the similar fundaments as the

Classic Theory. In this way it becomes possible not to rely on the other proposed theories, like micromorphic or multipolar theories. We believe that the Asymmetric Theory of Continuum opens enough possibilities to solve the main problems of the Classic Theory. In the classic elasticity the postulate of the Central Symmetry (Nowacki, 1986) argues against a direct use of rotations, e.g., strain rotation; however, we neglect the displacement motions in solid continuum (see considerations related to Eq. 4.6) and, thus, the related problems disappear. In the new approach the displacements do not enter into theory and therefore the Central Symmetry postulate does not work.

Thus, we may return to our proposed improvements of the classical approach, which leads to the asymmetric theories for solids and similarly for fluids.

Our presentation can be supported also by some thermodynamical considerations, especially those related to the vaporization processes: the thermodynamics of such vaporization effects may help to understand the turbulent anomalies influencing the transport motions. We should mention that in solids an influence of defects, or strictly speaking the defect density, becomes especially important close to fracture processes: the thermodynamics of the rebound dislocation motions helps to explain an appearance of fracture processes (Teisseyre 1985, 2001a, b; Teisseyre and Majewski 1990, 1995; Teisseyre and Górski 2006). A new approach including the strain rotation processes (Teisseyre and Majewski 2001; Teisseyre 2001a, b; Teisseyre 2011) may help to incorporate the related thermodynamical approach into a frame of the Asymmetric Theory of fluids. Instead of defect influence we consider some thermodynamical effects connected with the liquid–gas phase transition; such processes could influence the transport motions related to vortex structure and turbulent phenomena. The vaporization processes produce much bigger turbulent disturbances than the defect density accumulation at fracture events.

In this chapter we have focussed on our presentation of the elements of vortex theory based on the rotation transport mechanism. This is a new approach to vorticity and rotation transport (Teisseyre 2010) and the obtained relations describe the vortex motion and some related molecular bonds between the particles. According to the obtained solutions, the profiles of vortices, $\tilde{r}(z)$, may vary from almost linear, for very small $\alpha\,(\exp[-\alpha z] \approx 1 - \alpha z)$, up to very rapidly decreasing ones for the great αvalues. The vortex formation could effectively occur for the great Reynolds numbers above its critical value. The boundary conditions for a vortex appearance relate to the linear transport motions, but could be also affected by other neighbouring vortices; a row of such vortices might be formed. On the other hand, each vortex influences the linear flow encountering the vortices.

The theoretical approach presented here seems to be adequate for describing a model of turbulence mechanics based on the vortex dynamics. We hope that this new approach might open a way to considering a mechanical model of vortex and turbulence motions at the different thermodynamical conditions. Some further considerations on vortex motions and related molecular strains are discussed in Chap. 11.

References

Beirão da Veiga H (2007) Vorticity and regularity for viscous incompressible flows under the Dirichlet boundary condition. Results and related open problems. J Math Fluid Mech 9:506–516

Fan J, Jiang S, Nakamura G, Yong Zhou Y (2011) Logarithmically improved regularity criteria for the Navier–Stokes and MHD equations. J Math Fluid Mech 13(2011):557–571

Farwig R, Kozono H, Hermann Sohr H (2009) Energy-based regularity criteria for the Navier–Stokes equations. J Math Fluid Mech 11(2009):428–442

Landau LD, Lipshitz JM (1959) Fluid mechanics (Theoretical Physics, v. 6), (translated from Russian by Sykes JB, Reid WH). Pergamon Press, London, p 536

Landau LD, Lifszyc JM (2009) Hydrodynamika, (Hydrodynamics) Wydawnictwo Naukowe PWN Warszawa, (in Polish), p 671

Nowacki W (1986) Theory of asymmetric elasticity. PWN and Pergamon Press, Warszawa, p 383

Teisseyre R (1985) New earthquake rebound theory. Phys Earth Planet Inter 39:1–4

Teisseyre R (2001a) Shear band thermodynamic model of fracturing. In: Teisseyre R, Majewski E (eds) Earthquake thermodynamics and phase transformations in the Earth's interior, vol 76., International geophysics seriesAcademic Press, San Diego, pp 279–292

Teisseyre R (2001b) Evolution, propagation and diffusion of dislocation fields. In: Teisseyre R, Majewski E (eds) Earthquake thermodynamics and phase transformations in the earth's interior, vol 76., International Geophysical Series Academic Press, San Diego,San Francisco, NewYork, Boston, London, Sydney, Tokyo, pp 167–198

Teisseyre R (2008a) Introduction to asymmetric continuum: Dislocations in solids and extreme phenomena in fluids. Acta Geophys 56:259–269

Teisseyre R (2008b) Asymmetric continuum: standard theory. In: Teisseyre R, Nagahama H, Majewski E (eds) Physics of asymmetric continua : extreme and fracture processes. Springer, Berlin, pp 95–109

Teisseyre R (2009) Tutorial on new development in physics of rotation motions. Bull. Seismol.Soc. Am 99(2B):1028–1039

Teisseyre R (2010) Fluid theory with asymmetric molecular stresses: difference between vorticity and spin equations. Acta Geophys 58(6):1056–1071

Teisseyre R (2011) Why rotation seismology: Confrontation between classic and asymmetric theories. Bull Seismol Soc Am 101(4):1683–1691

Teisseyre R, Górski M (2009a) Fundamental deformations in asymmetric continuum: Motions and fracturing. Bull Seismol Soc Am 99(2B):1028–1039

Teisseyre R, Górski M (2009b) Transport in fracture processes: fragmentation and slip. Acta Geophys 57(3):583–599

Teisseyre R, Górski M, Teisseyre KP (2006) Fracture-band geometry and rotation energy release. In: Teisseyre R, Takeo M, Majewski E (eds) Earthquake source asymmetry,structural media and rotation effects. Springer, Berlin, pp 169–184

Teisseyre R, Majewski E (1990) Thermodynamics of line defects and earthquake processes. Acta Geophys Pol 38(4):355–373

Teisseyre R, Majewski E (1995) Earthquake thermodynamics. In: Teisseyre R (ed) Theory of earthquake premonitory and fracture processes. PWN, Warszawa, pp 586–590

Teisseyre R, Majewski E (2001) Thermodynamics of line defects and earthquake thermodynamics. In: Teisseyre R, Majewski E (eds) Earthquake thermodynamics and phase transformations in the Earth's interior, vol 76., International geophysics series Academic Press, San Diego, pp 261–278

Chapter 5
Defect Densities

5.1 Introduction

In solids an important role in the fracture processes is played by the defect densities; the defects reorganize the applied stress load and may form a special kind of defect-related anisotropy. There appears an important influence of the defect densities on the formation of micro-fractures and fragmentation processes. From the solution for edge dislocations we learn about the necessity to account for an asymmetry of stresses and strains. The defects and the induced stresses appear also due to the human activity, e.g., in the mining regions and water dam areas. In the theory of dislocations and dislocation arrays there appear relations between the defect densities and strains. The defect densities relate to the derivatives of strain fields; we present the appropriate relations for the asymmetric effects introduced in the frame of our Asymmetric Theory. Due to an influence of the dislocation density distribution, related to processes appearing at the extrema of densities, we may approach some micro-fracture processes. For the shear and rotation-induced fields we introduced additional conditions expressing a possible decreasing effect of the applied load due to some load vibration, near a resonance frequency related to the given material properties. A problem of resonance frequency can be considered per analogy to a classical problem of the black body radiation, it means, for radiation being near a thermodynamic equilibrium within a given material cavity. We also introduce some defect-like fields in a fluid continuum; here the defect-related elements may appear as some gas elements in a liquid and reversely the liquid intrusions in a gas continuum. We derive the related equations basing on a new approach with a definition of the molecular complex vector. For such a purpose, we define some circular defects in fluids as related to the vortex process and based on a molecular defect density and molecular dislocations.

R. Teisseyre and M. Teisseyre-Jeleńska, *Asymmetric Continuum*,
GeoPlanet: Earth and Planetary Sciences, DOI: 10.1007/978-3-642-31860-3_5,

5.2 Defect Densities

In solids we may arrive at the fracture formation criterion. An important role in the Asymmetric Continuum Theory is played by the defect densities, which may reorganize the applied stress load and even form a special kind of defect-related anisotropy. Generalization of the Peach-Koehler forces acting on the defects permits to define the induced strains which influence the effect of the applied stress load. The defect density field could also help to understand the formation of micro-fractures and fragmentation processes.

Solution for an edge dislocation presents some asymmetry in the strain components in the plane perpendicular to the dislocation line (wedge line); for a continuous distribution of dislocations this fact leads to a confrontation between the symmetry of shear strains in the classic continuum theory and an asymmetry of stresses related to the edge dislocations. Hence, there appears a problem how to express a direct relation between the distribution of the edge dislocations and the symmetric strains; however, this problem disappears when we consider a relation between the edge dislocations and the asymmetric strains.

Induced strains play also an important role in mining regions and water dam areas. We may consider a mechanism of reorganization of the applied stress load by some changes in rock-body defect distribution caused by human activity. A defect content increases with increasing stress load and related deformations; hence, there could appear a relation between seismic risk and deformation level. In particular, we may consider an increase of internal defects due to the mining works which reorganize also an internal stress distribution.

We should note that the strains resulting from the above-mentioned human activity may essentially differ from the existing load. In the case of the axial load, this approach helps to explain the formation of shear or rotational micro-fractures, usually recognized as fragmentation and slip motions.

Some material micro-destructions preceding the earthquakes or induced seismic events relate to very complicated nonlinear processes. The appearing micro-fractures may have a quite different character than that related to an external load; this is due to the role of internal defects existing as an integral element of any natural material. Therefore, in order to better understand the micro-destruction processes before any seismic event, we should take into account a distribution and influence of the different defect densities.

We should refer, on the one hand, to the boundary or initial conditions as described by the axial, deviatoric (shear) and antisymmetric (moment-related) stresses, and on the other hand to the unknown distribution of the internal defects and its changes due to human activity.

We first recall that the strain fields (axial, deviatoric and antisymmetric) can be presented by the adequate derivatives of some reference displacements, e.g., $\bar{u}$, $\hat{u}$, $\breve{u}$ (see Chap. 2). The real displacements appear only in fracture processes, while in a continuum the displacements have a mathematical sense. The independent strains represent the real physical fields.

In the theory of the dislocations and dislocation arrays (Eshelby et al. 1951, cf., Rybicki 1986) there appear relations between the defect densities and strains; the defect densities relate to the derivatives of strain fields (cf., Kossecka and DeWitt 1977). We will partly follow the paper by Teisseyre (2008a, b), in which there is the following definition of B_l, i.e., the complex Burgers and Frank vector (combination of the Burgers and Frank vectors)

$$B_l = \oint \left[\delta_{pl}\bar{E} + \hat{E}_{(pl)} + \breve{E}_{[pl]}\right] dx_p = \varepsilon_{qns} \iint \frac{\partial}{\partial x_n} \left[\delta_{sl}\bar{E} + \hat{E}_{(sl)} + \breve{E}_{[sl]}\right] ds_q$$
$$\text{or} \qquad B_l = \oint E_{pl} dx_p = \varepsilon_{qns} \iint \frac{\partial E_{sl}}{\partial x_n} ds_q \tag{5.1a}$$

where the strain fields are defined as follows

$$E_{sl} = \delta_{pl}\bar{E} + \hat{E}_{(sl)} + \breve{E}_{[sl]}; \bar{E} = \frac{1}{3}\sum_s E_{ss}$$
$$\hat{E}_{(sl)} = E_{(sl)} - \delta_{sl}\frac{1}{3}\sum_s E_{ss}, \sum_s \hat{E}_{(ss)} = 0 \tag{5.1b}$$

On the other hand, we present the corresponding definition based on a defect density as follows:

$$B_l = \iint \left(\alpha_{ql}^T - \frac{1}{2}\delta_{ql}\sum_i \alpha_{ii}^T\right) ds_q$$
$$B_l = \varepsilon_{qns} \iint \frac{sl}{\partial x_n} ds_q = \oint E_{sl} dx_s \tag{5.2a}$$

where the total dislocation field could split into the proper dislocations and disclinations: $\alpha_{pl}^T = \alpha_{pl} + \theta_{pl}$.

This definition can be related separately to the different types of densities

$$B_l = \iint \left[\left(\bar{\alpha}_{pl} - \frac{1}{2}\delta_{pl}\sum_i \bar{\alpha}_{ii}\right) + \hat{\alpha}_{pl} + \breve{\alpha}_{pl}\right] ds_p \tag{5.2b}$$

where we have introduced the screw, $\bar{\alpha}_{pl} - \frac{1}{2}\delta_{pl}\sum_i \bar{\alpha}_{ii}$, and edge dislocation density ($\hat{\alpha}_{pl}$), and moreover the disclination density ($\alpha_{pl}{}' \equiv \theta_{pl}$).

These complex deformations related to strain derivatives (Eq. 5.1), and direct defect influence (Eq. 5.2), present a total integrated deformation in two different but corresponding forms, known as the complex Burgers and Frank vector. In this way these two vector forms directly join the defect densities and strain derivatives (Teisseyre 2008a, b; cf., Kossecka and DeWitt 1977):

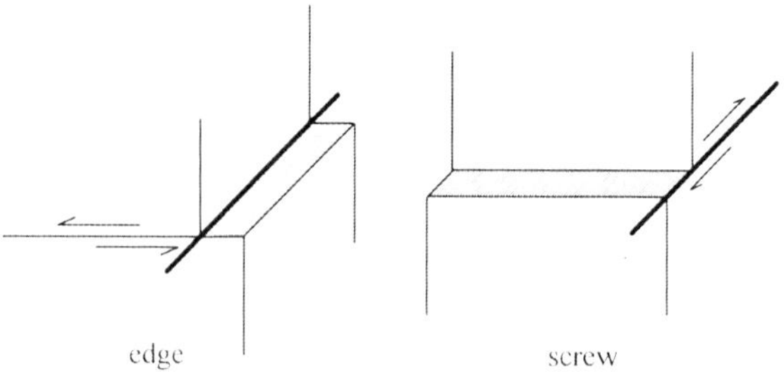

Fig. 5.1 The edge and screw dislocations; the heavy line denotes the dislocation wedge line

$$\alpha_{ql} - \frac{1}{2}\delta_{ql}\sum_i \alpha_{ii} = \varepsilon_{qns}\frac{\partial E_{sl}}{\partial x_n}$$

$$\alpha_{pl} - \frac{\delta_{pl}}{2}\sum_i \alpha_{ii}=\varepsilon_{pns}\left[\frac{\partial}{\partial x_n}\left(\bar{E}_{sl} - \frac{\delta_{sl}}{3}\sum_i \bar{E}_{ii}\right) + \frac{\partial}{\partial x_n}\hat{E}_{(sl)} + \frac{\partial}{\partial x_n}\breve{E}_{[sl]}\right] \tag{5.3a}$$

This relation can be split into ones for edge, $\bar{\alpha}_{pl}$, and screw, $\hat{\alpha}_{pl}$, dislocation densities and for disclination, $\breve{\alpha}_{pl}$, density:

$$\bar{\alpha}_{pl} - \frac{1}{2}\delta_{ps}\sum_s \bar{\alpha}_{sl}=\varepsilon_{pns}\frac{\partial}{\partial x_n}\left(\bar{E}_{sl} - \frac{\delta_{sl}}{3}\sum_i \bar{E}_{ii}\right)=\xi^0\varepsilon_{pns}\frac{\delta_{sl}\partial}{\partial x_n}\left(\frac{\partial u_l}{\partial x_s} - \frac{1}{3}\sum_i\frac{\partial u_i}{\partial x_i}\right)$$

$$\hat{\alpha}_{pl} - \frac{1}{2}\delta_{pl}\sum_i \hat{\alpha}_{ii} = \varepsilon_{pns}\frac{\partial}{\partial x_n}\hat{E}_{(sl)} = e^0\varepsilon_{pns}\frac{\partial}{\partial x_n}\left(\frac{1}{2}\frac{\partial u_s}{\partial x_l} + \frac{1}{2}\frac{\partial u_l}{2\partial x_s} - \frac{\delta_{sl}}{3}\sum_i\frac{\partial u_i}{\partial x_i}\right)$$

$$\breve{\alpha}_{pl} = \theta_{pl} = \varepsilon_{pns}\frac{\partial}{\partial x_n}\breve{E}_{[sl]} = \chi^0\varepsilon_{pns}\frac{\partial}{\partial x_n}\left(\frac{1}{2}\frac{\partial u_s}{\partial x_l} - \frac{1}{2}\frac{\partial u_l}{\partial x_s}\right) \tag{5.3b}$$

In particular, for the disclination density we may also write:

$$\breve{\alpha}_{pl} = \theta_{pl};\ \theta_{pl} = \varepsilon_{pns}\frac{\partial}{\partial x_n}\breve{E}_{[sl]};\ \theta_{pp} = \varepsilon_{pns}\frac{\partial}{\partial x_n}\breve{E}_{[sp]};\ \sum_p \theta_{pp} = 0 \tag{5.3c}$$

In these relations (Eq. 5.3a–c), we have also introduced the reference displacement fields used to define mathematically the expressions for strain distribution and possible phase shifts between these independent strains; this independence means that these fields could be released, e.g., in a seismic source, as the completely independent or interdependent fields (in this last case, related to the mutually dependent process with a possible phase shift between the released motions; cf., Chap. 2) (Fig. 5.1).

In the fracture processes the dislocations can form arrays leading to cracks, while the disclination arrays may help to form the micro-vortex defects or fragmentations and cracks. Moreover, the arrays formed by the defects could play an important role in the induced seismicity problems. As mentioned, some dislocation density distributions with the extrema of densities can lead to fracture (or microfracture)

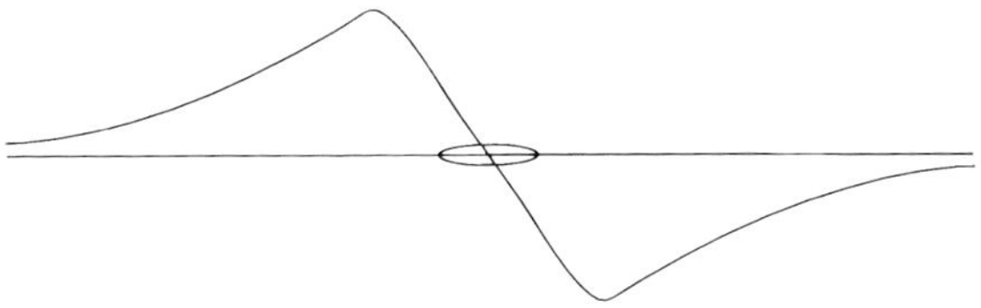

Fig. 5.2 The two extremes of the opposite dislocation densities present a possible effect related to the two neighboring cracks; in this figure, near the center, these two cracks join each other, leading to an annihilation process and a related energy release

processes; as visualized in Fig. 5.2. In such a way, seismic risk may increase due to an accumulation of defects.

The relations for the axial-strain, shear-strain and rotation-strain (cf., Eq. 5.3b), can be presented also in relation to stresses as follows:

$$\alpha_{pl} - \frac{\delta_{pl}}{2}\alpha_{ss} = \varepsilon_{pmk}\frac{\partial E_{kl}}{\partial x_m} = \varepsilon_{pmk}\left(\delta_{kl}\frac{\partial \bar{E}}{\partial x_m} + \frac{\partial \hat{E}_{(kl)}}{\partial x_m}\right) = \frac{\varepsilon_{pmk}}{2\mu}\frac{\partial\left(S_{(kl)} - \frac{v}{1+v}\delta_{kl}S_{ss}\right)}{\partial x_m} \tag{5.4a}$$

where v is the Poisson ratio.

$$\theta_{pl} = \varepsilon_{pns}\frac{\partial}{\partial x_n}E_{[sl]} = \frac{\varepsilon_{pmk}}{2\mu}\frac{\partial S_{[kl]}}{\partial x_m} \tag{5.4b}$$

A special defect concentration may lead, e.g., to an array of the edge dislocations $\Sigma\hat{\alpha}_{12}$ (on a plane $x_3 = 0$)and the related extreme stress concentration may lead to a crack formation.

The presented relations between the strains and defect densities can be considered when discussing the compatibility tensors (Kossecka and DeWitt 1977); now we may define these tensors considering both the symmetric and antisymmetric parts (Boratynski and Teisseyre 2006; Teisseyre 2008a, b):

$$\begin{aligned} I_{(ij)} &= \varepsilon_{ikm}\varepsilon_{jtn}\frac{\partial^2 E_{(mn)}}{\partial x_k \partial x_t} = \varepsilon_{ikm}\varepsilon_{jtn}\frac{\partial^2 \hat{E}_{(mn)}}{\partial x_k \partial x_t} = 0; \\ I_{[ij]} &= \varepsilon_{ikm}\varepsilon_{jtn}\frac{\partial E_{[kl]}}{\partial x_k \partial x_t} = \varepsilon_{ikm}\varepsilon_{jtn}\frac{\partial E_{[kl]}}{\partial x_k \partial x_t} = 0 \end{aligned} \tag{5.5a}$$

When we express these tensors by the dislocation densities (Kossecka and DeWitt 1977) we get:

$$\begin{aligned} \alpha_{pl} - \frac{\delta_{pl}}{2}\alpha_{ss} &= \varepsilon_{pmk}\frac{\partial E_{kl}}{\partial x_m} \\ I_{pq} &= \varepsilon_{pmk}\frac{\partial \alpha_{qk}}{\partial x_m} \end{aligned} \tag{5.5b}$$

and $\alpha_{jn} - \frac{\delta_{jn}}{2}\sum_{s}\alpha_{ss} = \varepsilon_{jtm}\frac{\partial E_{(mn)}}{\partial x_t}$ and $I_{pq} = \varepsilon_{pmk}\frac{\partial \alpha_{qk}}{\partial x_m}$ we obtain:

$$I_{(ij)} = \varepsilon_{ikn}\varepsilon_{jtm}\frac{\partial^2 E_{(mn)}}{\partial x_k \partial x_t} = \varepsilon_{ikn}\frac{\partial \alpha_{jn} - \frac{\delta_{jn}}{2}\sum_{s}\alpha_{ss}}{\partial x_k} = \varepsilon_{ikn}\frac{\partial \alpha_{jn}}{\partial x_k} = 0$$
$$I_{[ij]} = \varepsilon_{ikn}\varepsilon_{jtm}\frac{\partial^2 E_{[mn]}}{\partial x_k \partial x_t} = \varepsilon_{ikn}\frac{\partial \alpha_{jn} - \frac{\delta_{jn}}{2}\sum_{s}\alpha_{ss}}{\partial x_k} = \varepsilon_{ikn}\frac{\partial \alpha_{jn}}{\partial x_k} + \varepsilon_{ijk}\frac{\sum_{s}\alpha_{ss}}{2\partial x_k} = 0 \tag{5.6}$$

However, when the conditions for strains and for dislocation densities (Eq. 5.5) become not fulfilled, then we may understand that this phenomenon indicates a possible precursor of fracture process; thus, a deviation of the expressions (5.5) from zero may serve as the fracture formation criterion. In this way we approach the backgrounds of the fracture criteria; we will return to these problems in Chap. 9 related to the electro-magnetic effects. Here, we explain that such deviations could be related to the mutual annihilation or formation the oppositely oriented dislocations; we may call them the positive and negative dislocations (arbitrarily defined). The dislocations can form arrays leading to cracks, while the disclination arrays defined above (related to gradient of rotations) will form the micro-vortex defects or fragmentation-cracks. An important problem relates to the defect and crack flow velocity; according to Holländer (1960), and Teisseyre (1980, 1985), we can write for defects and their flow velocity:

$$\frac{\partial}{\partial t}\alpha_{kl} + \frac{\partial}{\partial x_n}(\alpha_{kn}\upsilon_l - \alpha_{kl}\upsilon_n) = 0 \tag{5.7}$$

Starting from this continuity relation we may take into account the oppositely oriented dislocations, the 'positive', α_{kl}, and 'negative', β_{kl}, ones; in such a case we should include a possibility of creation, $[\alpha\beta]$, or mutual annihilation, $[\beta\alpha]$ of the oppositely oriented dislocations and we may write, after Teisseyre (1990a, b), a more complicated relation including these creation and annihilation processes. However, it seems that the continuity relation (Eq. 5.7) should include also such additional rapid creation and annihilation processes.

We should mention that the circular defects, e.g., loop dislocations, differ from the disclinations; the first mean the defect circular forms for the dislocation lines, while the other mean a deformation caused by an angular shift along a straight dislocation line. Following relations (5.1–5.7) derived for the dislocations, we may construct the circular defects.

The fracture processes depend, of course, on the defect densities. Here, an important contribution is the formula by Eshelby et al. (1951), stating that the stress concentration at a tip of a linear dislocation array leads to multiplication of the applied stress load by a number of dislocations in an array, n:

$$S = nS^{LOAD} \tag{5.8}$$

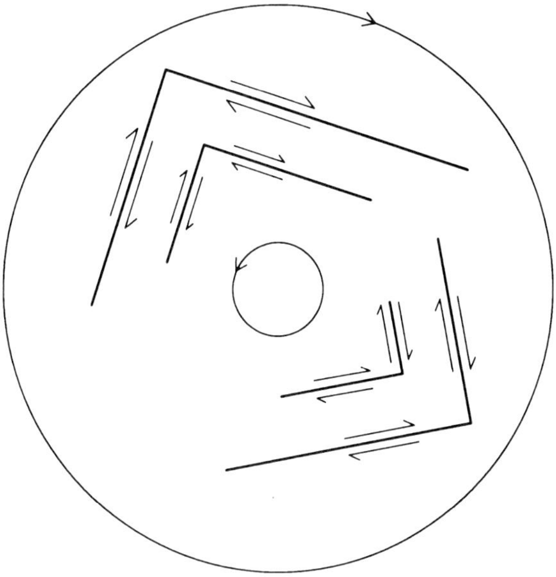

Fig. 5.3 Shear fractures and fragmentation; due to confining load, shears of opposite-orientation along the perpendicular planes may be formed together with an opposite-sense rotation

This is a very important relation permitting to define the local stress concentration due to a defect distribution. An applied stress load acts on defects and may lead to rearrangement of defect distribution and to the formation of defect arrays. In the continuum theory the defect arrays can be described by an increase of defect densities along some direction, according to material parameters and existing defect distribution. Further on, in the earthquake processes the external load, that is, the given external symmetric stresses and antisymmetric stresses (stress moments), can lead to the fracture processes. We should note that an axial strain, even very strong, does not lead directly to fracture processes, while due the presence of defects there appear some induced local strains of shear and rotation type. In a neighbourhood of such a process there may appear induced strains of similar types, but of opposite orientation in order to compensate for these induced strain anomalies. Thus, due to these induced strains, the shear fracture and fragmentation processes may be excited, meaning the concentric rotational defects (cf., Fig. 5.3, Teisseyre and Górski 2012).

Some other induced effects related to the confining load are presented in Fig. 5.4.

We might suppose also that the transport of rotation motions becomes possible only in the granulated structures or in those undergoing micro-fragmentation processes.

We present also the disclination line related to the induced rotational deformation (Fig. 5.5).

An extreme example of molecular rotational deformation in solids might even look similar to a vortex motion (Fig. 5.6).

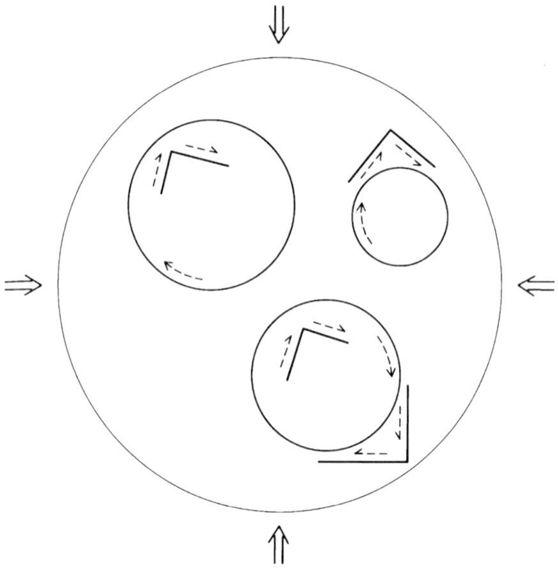

Fig. 5.4 Fragmentation centers and induced opposite-orientation shears related to a confining load

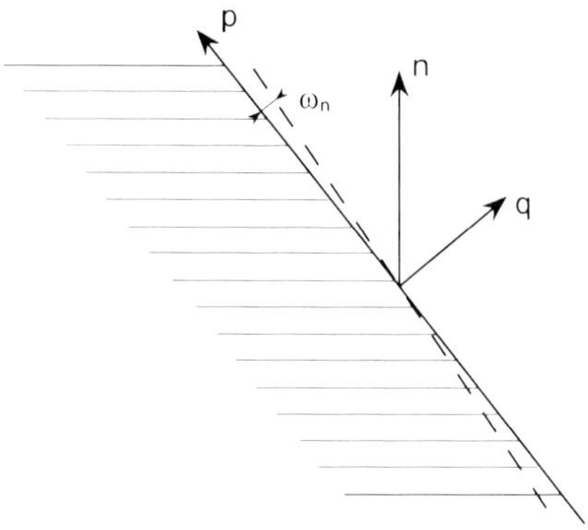

Fig. 5.5 Disclination line due to the induced rotational deformation

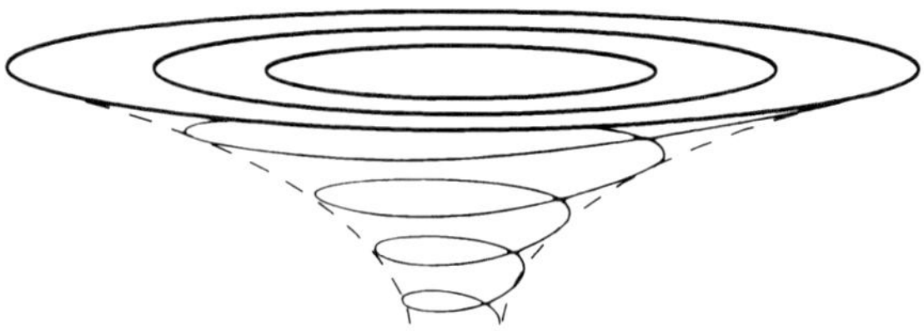

Fig. 5.6 A micro-rotational molecular deformation

We believe that these types of processes, appearing due to some internal defect contents, especially a possible formation of an array of defects under a given constant load, could lead to a great defect concentration; a number of elementary linear defects (here we neglect the point defects) can increase with time and a fatigue process may appear. Thus, we may expect that the dislocation processes will lead us to the micro-crack phenomena and finally to crack formation and a local fracture.

In Chap. 7 we will discuss a problem of the induced stresses due to the defect concentration; here we may just note that an extreme defect concentration could break the relations between the concentrated dislocations and related stress derivatives which are presented here (cf., Eq. 5.4). As a consequence of such breaks, the compatibility between strains and defects might be violated and a fracture can appear.

Some consequence of the material micro-destructions preceding the earthquakes or induced seismic events relate to very complicated, surely nonlinear processes. A character of such micro-fractures may be quite different from that related to the existing external load; this is due to the role of internal defects existing as an integral element of any natural material. Therefore, to understand better the forwarding micro-destruction processes before any seismic event, we need to account for a distribution of different defect densities.

We should refer, on the one side, to the boundary or initial conditions as a combination of the axial (compressional), deviatoric (shear) and antisymmetric (moment related) stresses, and, on the other side, to the unknown combination of the internal defect distribution.

Due to a given distribution of defects, we may estimate the external forces induced by the given external stress fields, a given compression, deviatoric shear and rotation stress, $\bar{S}^{Ext}, \hat{S}^{Ext}_{(sk)}, \breve{S}^{Ext}_{[sk]}$, forming the external forces:

$$\begin{aligned} \bar{F}_n &= \varepsilon_{nsq}\bar{S}^{Ext}\delta_{sk}b_k v_q = \varepsilon_{nkq}\bar{S}^{Ext}\alpha_{qk}, \quad \hat{F}_n = \varepsilon_{nsq}\hat{S}^{Ext}_{(sk)}b_k v_q = \varepsilon_{nsq}\hat{S}^{Ext}_{(sk)}\alpha_{qk} \\ \breve{F}_n &\to M_n = \varepsilon_{nsq}\breve{S}^{Ext}_{[sk]}b_k v_q = \varepsilon_{nsq}\breve{S}^{Ext}_{[sk]}\alpha_{qk}, \end{aligned} \tag{5.9}$$

where $\bar{S} = \frac{1}{3}S_{ii}$ and $b_k v_q = \alpha_{qk}$.

For the shear and rotation-induced fields we may introduce additional conditions meaning a possible decreasing effect of the applied load due to the load vibration, with ν_0 expressing a resonance frequency related to given material properties. A problem of resonance frequency can be considered in terms of the classical formulation of the black body radiation, it means, for radiation being in thermodynamic equilibrium within a given material cavity. Such a radiated energy per wave length unit may be described by the Planck relation (cf. Chap. 10).

Thus, a strength may be assumed to be decreasing due to the frequency related effect; such an effect may reduce the strength, from the static value down; thus, under vibration load a strength may reach its minimum while the deformation reaches its maximum (Chelidze et al. 2010).

Following this approach, we assume that the energy break under a given confining load and shears will depend on the frequency of the applied additional load. At an end of this consideration we repeat once again the important conclusion that in the mining and water infiltrated areas an induced seismicity may appear due to a great concentration of material defects, mining corridors, and also due to water infiltration.

In fluids we may also introduce some defect-like fields; in a liquid continuum some gas elements may play a role corresponding to defects and reversely in a gas continuum such a role may be played by the liquid intrusions.

As examples we may mention that when the water approaches boiling, the appearance of vapor elements could be treated similarly to the defects in solids discussed above; reversely, rain drops in the atmosphere might be considered as the defect elements.

We may follow the considerations related to defects in solids: instead of Eq. (5.1) we define the molecular complex vector. However, the circular defects appearing in fluids relate to the vortex process and therefore we will use here the cylindrical coordinate system with the defect lines along the circles perpendicular to the z-direction; we may write, after (5.1–5.4):

$$\tilde{B}_l = \oint \tilde{E}_{\varphi l} r d\varphi = \oint \left[\delta_{\varphi l}\bar{\bar{E}} + \hat{\tilde{E}}_{(\varphi l)} + \bar{\tilde{E}}_{[\varphi l]}\right] r d\varphi = \varepsilon_{zns} \iint \frac{\partial}{\partial x_n} \tilde{E}_{sl} ds_z \quad (5.10)$$

where the molecular strain fields, $\tilde{E}_{sl}$, and related equations of motion follow those presented in Chap. 2.

On other hand, we present the corresponding definition based on a molecular defect density and molecular dislocations (cf., Eq. 5.2):

$$\tilde{B}_s = \iint \left(\tilde{\alpha}_{ps} - \frac{1}{2}\delta_{ps}\sum_i \tilde{\alpha}_{ii}\right) ds_p \quad (5.11)$$

When applying these relations for $\tilde{\alpha}_{\varphi s} = \left\{\tilde{\alpha}_{\varphi r}, \tilde{\alpha}_{\varphi\varphi}, \tilde{\alpha}_{\varphi z}\right\}$, we obtain a direct expression for a loop disclination, $\tilde{\alpha}_{\varphi\varphi}$.

The total molecular dislocation will be given as:

$$\tilde{\alpha}_{ps} - \frac{1}{2}\delta_{ps}\sum_i \tilde{\alpha}_{ii} = \varepsilon_{pns}\frac{\partial}{\partial x_n}\tilde{E}_{sl},\ \tilde{\alpha}_{pl}^T = \varepsilon_{pns}\frac{\partial}{\partial x_n}\tilde{E}_{sl} \quad (5.12)$$

$$\tilde{B}_s = \iint \left(\tilde{\alpha}_{ps} - \frac{1}{2}\delta_{ps}\sum_i \tilde{\alpha}_{ii}\right) ds_p \quad (5.13)$$

Thus, the complex deformations (5.12) lead us to the relation:

$$\begin{aligned} \tilde{\alpha}_{\varphi l} - \frac{1}{2}\delta_{\varphi l}\sum_i \tilde{\alpha}_{ii} &= \varepsilon_{\varphi ns}\frac{\partial}{n}\tilde{E}_{sl} \\ \tilde{\alpha}_{\varphi l} - \frac{1}{2}\delta_{\varphi l}\sum_i \tilde{\alpha}_{ii} &= \varepsilon_{\varphi ns}\left[\frac{\partial}{\partial x_n}\delta_{\varphi l}\bar{\bar{E}} + \hat{\tilde{E}}_{(\varphi l)} + \bar{\tilde{E}}_{[]}\right] \end{aligned} \quad (5.14)$$

For an array of edge dislocations on plane we may arrive at a circular crack.

In this consideration, relations for the density (cf., Eqs. 3.5 and 2.13) play an important role; we may propose important relations with the complex float transport including external float and also the density variations, causing the material volume changes, especially important for a gas medium:

$$\rho\frac{\partial(\rho\upsilon_i)}{\partial t}=\upsilon_i\frac{\partial\rho}{\partial t}+\rho\frac{\partial\upsilon_i}{\partial t}+\upsilon_i\sum_n(\upsilon_n+\bar{\upsilon}_n)\frac{\partial\rho}{\partial x_n}+\rho\sum_s(\upsilon_s+\bar{\upsilon}_s)\frac{\partial\upsilon_i}{\partial x_s}$$
$$=\eta\sum_s\frac{\partial^2\upsilon_i}{\partial x_s^2}+F_i \tag{5.15}$$

$$\frac{d\rho}{dt}=\frac{\partial\rho}{\partial t}+\sum_n(\upsilon_n+\bar{\upsilon}_n)\frac{\partial\rho}{\partial x_n}+\rho\sum_n\frac{\partial\upsilon_n}{\partial x_n}=0 \tag{5.16}$$

or, when combining these two relations, we obtain the complex one:

$$\rho\frac{\partial\upsilon_i}{\partial t}-\upsilon_i\sum_n(\upsilon_n+\bar{\upsilon}_n)\frac{\partial\rho}{\partial x_n}-\rho\upsilon_i\sum_n\frac{\partial\upsilon_n}{\partial x_n}+\sum_s(\upsilon_s+\bar{\upsilon}_s)\left[\upsilon_i\frac{\partial\rho}{\partial x_s}+\rho\frac{\partial\upsilon_i}{\partial x_s}\right]+$$
$$\rho\frac{\partial\upsilon_i}{\partial x_s}=\eta\sum_s\frac{\partial^2\upsilon_i}{\partial x_s^2}+F_i \tag{5.17}$$

We may relate the fluid molecular strains with some transport fields and with a molecular defect density field (5.16). We can put for the molecular strains:

$$\tilde{E}_{(sl)}=\frac{1}{2}\left(\frac{\partial\bar{\upsilon}_l}{\partial x_s}+\frac{\partial\bar{\upsilon}_s}{\partial x_l}\right),\ \tilde{E}_{[sl]}=\frac{1}{2}\left(\frac{\partial\bar{\upsilon}_l}{\partial x_s}-\frac{\partial\bar{\upsilon}_s}{\partial x_l}\right);$$
$$\tilde{E}_{sl}=\tilde{E}_{(sl)}+\tilde{E}_{[sl]} \tag{5.18a}$$

and then we obtain for defect densities:

$$\tilde{\alpha}_{pl}=\varepsilon_{pns}\frac{\partial}{\partial x_n}\tilde{E}_{sl};\ \tilde{\alpha}_{pl}=\varepsilon_{pns}\frac{1}{2}\frac{\partial^2\bar{\upsilon}_s}{\partial x_n\partial x_l},\ \text{ where } \tilde{\alpha}_{pp}=0 \tag{5.18b}$$

where we have presented the molecular strain with the help of some float field, $\bar{\upsilon}_s$.

Further, we obtain

$$\tilde{\alpha}_{ps}=\varepsilon_{pns}\frac{\partial\tilde{E}_{ss}}{\partial x_n};\ \tilde{\alpha}_{ps}=\varepsilon_{pns}\frac{1}{2}\frac{\partial^2\bar{\upsilon}_s}{\partial x_n\partial x_s}\ \rightarrow\ \tilde{\alpha}_{pp}=0$$
$$\tilde{\alpha}_{pp+1}=-\frac{1}{2}\frac{\partial^2\bar{\upsilon}_{p+1}}{\partial x_{p+2}\partial x_{p+1}},\ \tilde{\alpha}_{pp+2}=\frac{\partial^2\bar{\upsilon}_{p+2}}{\partial x_{p+1}\partial x_{p+2}} \tag{5.18c}$$

These defects appear due to the continuum under flow, and for the pressure-related case, $\tilde{E}_{sl}\rightarrow\tilde{E}_{ss}=-\frac{1}{2\eta}\frac{\partial p}{\partial t}$ and $\frac{1}{3}\sum_s\tilde{E}_{ss}=-\frac{1}{2\eta}\frac{\partial p}{\partial t}$. Thus, we can write:

$$\tilde{\alpha}_{pp+1}=-\frac{1}{2\eta}\frac{\partial}{\partial x_{p+2}}\frac{\partial p}{\partial t};\ \tilde{\alpha}_{pp+2}=\frac{1}{2\eta}\frac{\partial}{\partial x_{p+1}}\frac{\partial p}{\partial t} \tag{5.19c}$$

or in the cylindrical coordinates

$$\begin{aligned}\tilde{\alpha}_{r\varphi} &= -\frac{1}{2\eta}\frac{\partial}{\partial z}\frac{\partial p}{\partial t};\ \tilde{\alpha}_{rz} = \frac{1}{2\eta}\frac{\partial}{r\partial\varphi}\frac{\partial p}{\partial t}\\ \tilde{\alpha}_{\varphi z} &= -\frac{1}{2\eta}\frac{\partial}{\partial r}\frac{\partial p}{\partial t};\ \tilde{\alpha}_{\varphi r} = \frac{1}{2\eta}\frac{\partial}{\partial z}\frac{\partial p}{\partial t}\\ \tilde{\alpha}_{zr} &= -\frac{1}{2\eta}\frac{\partial}{r\partial\varphi}\frac{\partial p}{\partial t};\ \tilde{\alpha}_{z\varphi} = \frac{1}{2\eta}\frac{\partial}{\partial r}\frac{\partial p}{\partial t}\end{aligned} \tag{5.19d}$$

Here, we should notice a possible lengthening and shortening effects of the pressure derivatives; we note that the appearance of these molecular defect deformations becomes signalized by the sonar effects, $\frac{\partial p}{\partial t}$ (cf., Eq. 5.19).

References

Boratyński W, Teisseyre R (2006) Continuum with rotation nuclei and defects: dislocations and disclination fields. In: Teisseyre R, Takeo M, Majewski E (eds) Earthquake source asymmetry, structural media and rotation effects. Springer, Berlin, pp 57–66

Chelidze T, Matcharashvili T, Lursmanashvili O, Varamashvili N, Zhukova E, Meparidze E (2010) Triggering and synchronization of stick-sip: experiments on spring-slider system. In: De Rubeis V, Czechowski Zb, Teisseyre R (eds) Synchronization and triggering: from fracture to earthquake processes. Springer, Berlin, pp 123–164

Eshelby JD, Frank FC, Nabarro FRN (1951) The equilibrium of linear arrays of dislocations. Philos Mag 42:351–364

Holländer EF (1960) The basic equations of the continuous distribution of dislocations, I,II,III. Czech J Phys B 10:409–418 479–487, 551–558

Kossecka, E, DeWitt R (1977) Disclination kinematic. Arch Mech 29:633–651

Rybicki K (1986) Dislocations and their geophysical applications. In: Teisseyre R (ed) Continuum theories in solid earth physics. Elsevier-PWN, Amsterdam-Warszawa, pp 18–186

Teisseyre R (1980) Earthquake premonitory sequence – dislocation processes and fracturing. Boll Geofis Teor Appl 22:245–254

Teisseyre R (1985) New earthquake rebound theory. Phys Earth Planet Inter 39:1–4

Teisseyre R (1990a) Earthquake rebound: energy release and defect density drop. Acta Geophys Pol XXXVIII(1):15–20

Teisseyre R (1990b) Earthquake premonitory and rebound theory: synthesis and revision of principles. Acta Geophys Pol XXXVIII(3):269–278

Teisseyre R (2008a) Introduction to asymmetric continuum: dislocations in solids and extreme phenomena in fluids. Acta Geophys 56:259–269

Teisseyre R (2008b) Asymmetric continuum: standard theory. In: Teisseyre R, Nagahama H, Majewski E (eds) Physics of asymmetric continua : extreme and fracture processes. Springer, Berlin, pp 95–109

Teisseyre R, Górski M (2012) Induced strains and defect continuum theory: internal reorganization of load. Acta Geophys 60(1):24–42

Chapter 6
Structural and Fracture Anisotropy

6.1 Introduction

The Asymmetric Continuum Theory may lead us to problems of the induced anisotropy, caused by the defect content, friction and fracture pattern. We consider such a problem of the anisotropy as reduced to the 2D anisotropy caused by an influence of some linear dislocation system. The fracture processes trace the pre-slip structures, which leads to the related anisotropic fracture ability and friction properties. We make an attempt to express the 2D anisotropy, as related to the antisymmetric part of stresses (stress moment anisotropy): for this problem we use the skew coordinate system. The fracture processes trace the pre-slip structure of media revealed in the formation mylonite layers leading to inertia-related anisotropy and related fracture ability and friction properties. To this aim, we have introduced a geometrical approach to the problem of skew and fracture anisotropy.

6.2 Structural and Fracture Anisotropy

An extension of the Asymmetric Continuum Theory may lead us to problems of the induced anisotropy, caused by the defect content, and related to friction anisotropy and fracture pattern. We should keep in mind that in the classic elastic continuum theory, any independent rotation motions are excluded; it means that there do not exist any constitutive laws relating an elastic response to the rotational deformations of bonds in a lattice network. However, we can record seismic rotation fields; some experimental data clearly indicate that the independent spin (rotation) motions become detectable. These rotation motions can be measured, among other methods, by means of ring laser or fibre optics interferometers, based on the Sagnac principle (Schreiber et al. 2006; Takeo 2006; Jaroszewicz et al. 2006) or by rotation seismographs (Moriya and Teisseyre 2006; Wiszniowski 2006).

These fields, spin (rotation strain) and twist (shear) motions, form together a propagation system (cf., Chaps. 1 and 2). The twist deformations present the grain

R. Teisseyre and M. Teisseyre-Jeleńska, *Asymmetric Continuum*,
GeoPlanet: Earth and Planetary Sciences, DOI: 10.1007/978-3-642-31860-3_6,

deformations caused by elastic strain; however, when considering the grains as "rigid" points of continuum, such a twist deformation converts to a kind of 3D space curvature (Teisseyre 2005; Teisseyre et al. 2006b).

We will consider the asymmetric continuum with the isotropic and anisotropic parts for strains and stresses:

$$E_{ns} = \bar{E}_{ns} + \tilde{E}_{ns} \text{ and } S_{ns} = \bar{S}_{ns} + \tilde{S}_{ns} \tag{6.1}$$

The balance equations for the isotropic fields can split into the parts related to symmetric and antisymmetric strains and stresses. The anisotropic fields, $\tilde{E}_{(ki)}$ and $\tilde{E}_{[ki]}$, cease to be independent; thus we can write:

$$\bar{E}_{ns} = \bar{E}_{(ns)} + \bar{E}_{[ns]}, \text{ but } \tilde{E}_{ns} \neq \tilde{E}_{(ns)} + \tilde{E}_{[ns]}, \text{ and thus } E_{ns} \neq E_{(ns)} + E_{[ns]} \tag{6.2a}$$

and similarly for stresses:

$$\bar{S}_{ns} = \bar{S}_{(ns)} + \bar{S}_{[ns]}, \text{ but } \tilde{S}_{ns} \neq \tilde{S}_{(ns)} + \tilde{S}'_{[ns]}, \text{ and thus } S_{ns} \neq S_{(ns)} + S_{[ns]} \tag{6.2b}$$

We may remark that the expressions (6.2a) are quite similar to the Kröner (1981) theory based on the total fields splitting into the elastic and self-fields (Kröner 1981; cf., Teisseyre 2005, 2006). We use this similarity only for the asymmetric fields, as in Eq. (6.2a). Of course, for the isotropic parts, $\bar{E}_{ns}$, the balance equations split into the parts related to the independent symmetric and antisymmetric strains, or stresses.

Recalling the structure of the Kröner theory (Kröner 1981), we may express the isotropic strains and stresses by a difference between physical and anisotropic fields (in comparison to the Kröner theory there is the following equivalence: elastic/physical ↔ isotropic, self ↔ anisotropic and total ↔ physical). Comparing our assumptions to the Kröner (1981) theory we state that the Kröner elastic fields correspond here to the isotropic fields and the self fields to the anisotropic fields, while the total fields correspond to the physical ones (in the Kröner theory the physical fields are represented by the elastic fields).

We assume that the isotropic stresses are related to physical strains through the ideal elasticity relation (cf., Chap. 2).

We might note that according to Shimbo (1975)??? the following constitutive law joins the antisymmetric isotropic stresses with the antisymmetric isotropic strains (see: Teisseyre et al. 2006, Chaps. 4–6):

$$\bar{S}_{[kn]} = 2\bar{\mu}\bar{E}_{[kn]}, \tag{6.3}$$

where $\bar{\mu}$ is an additional constant—the rotation rigidity.

In other chapters we have put in a first approximation: $\bar{\mu} \approx \mu$; however, here we assume that $\bar{\mu} \neq \mu$.

Returning to the problem of anisotropy we assume that it is reduced to the 2D anisotropy as being due to a possible influence of some linear dislocation field system (Teisseyre 2006); thus, the related anisotropy is expressed by the 2D skew coordinate system X^K ($K, S, \ldots. = \{1, 2\}$).

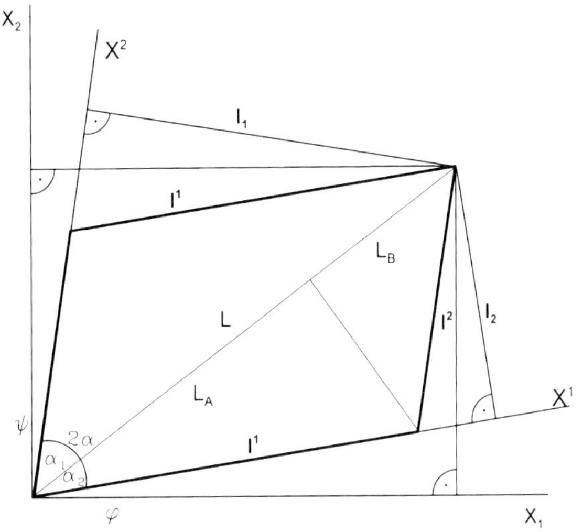

Fig. 6.1 The 2D internal friction anisotropy: a skew system of pre-slip planes

We write the constitutive law for the anisotropic part with a related material anisotropy by modification of the constitutive laws for the shear and rotation processes proceeding at different rates along the $X^1 = 0$, $X^2 = 0$ planes. We put:

$$\tilde{\boldsymbol{\mu}} = \tilde{\mu}_1 \mathbf{e}_1 + \tilde{\mu}_2 \mathbf{e}_2 \tag{6.4a}$$

and

$$\tilde{S}_{KS} = 2\tilde{\boldsymbol{\mu}}\tilde{E}_{KS} \tag{6.4b}$$

In the 2D skew coordinate system $\{X^K\}$, $K = 1, 2$ (Fig. 6.1), we recall the contravariant (parallel projection) and covariant (perpendicular projection) components of any vector **L**; with the help of metric tensor $g_{NS}(g_{11}, g_{22}, g_{12})$ we may put for a length element in that system:

$$L^2 = g_{11}l^1l^1 + g_{22}l^2l^2 + 2g_{12}l^1l^2 = l^1l^1\cos^2\alpha_1 + l^2l^2\cos^2\alpha_2 + 2l^1l^2\cos\alpha_1\cos\alpha_2 \tag{6.5}$$

and further, for the base vectors $\mathbf{e}_1$, $\mathbf{e}_2$ of the skew system:

$$g_{KS} = \mathrm{e}L\cos\alpha_1,\ l_2 = L\cos\alpha_2,\quad l^1 = {}^{l_1}\!/_{\cos 2\alpha},\quad l^2 = {}^{l_2}\!/_{\cos 2\alpha} \tag{6.6a}$$

$$l_1 = L\cos\alpha_1,\ l_2 = L\cos\alpha_2,\quad l^1 = {}^{l_1}\!/_{\cos 2\alpha},\quad l^2 = {}^{l_2}\!/_{\cos 2\alpha} \tag{6.6b}$$

$$\mathbf{L} = \mathbf{e}_1 l^1 + \mathbf{e}_2 l^2,\quad |\mathbf{e}_1| = \cos\alpha_1,\quad |\mathbf{e}_2| = \cos\alpha_2,\quad L = l^1\cos\alpha_1 + l^2\cos\alpha_2. \tag{6.6c}$$

Hence, we obtain

$$g_{KS} = \mathbf{e}_K \mathbf{e}_S; \quad g_{11} = \cos^2 \alpha_1, \quad g_{22} = \cos^2 \alpha_2, \quad g_{12} = \cos \alpha_1 \cos \alpha_2 \tag{6.7}$$

This approach becomes possible under the condition that our skew system is directly related to the considered anisotropy, and of course we may return to the rectangular system $\{x_k\} \equiv \{x_1, x_2\}$.

With these expressions we might follow the procedure usually applied for anisotropy, with the eikonal equation. However, instead of this contemporary approach to anisotropy based on the eikonal equation, we prefer to use a more direct geometrical approach.

We refer here to the paper by Teisseyre (1955) related to the problem of ray theory (see: Yajima and Nagahama 2006) and we may express the 2D anisotropic stresses using the related skew coordinate system. The fracture processes trace the pre-slip structures revealed in the formation of mylonite layers. The mylonite layer zones form an anisotropic system which leads to the related anisotropic fracture ability and friction properties. Figure 6.1 presents the skew system of the privileged slip planes having different ability for slip motions. The induced anisotropy might be related also to the different radii of grains.

We should underline again that the anisotropic parts, in the 2D (X, Y) anisotropy, cannot be divided into the separate symmetric and antisymmetric parts; however, these anisotropic fields form two different groups, $\tilde{E}_{12}$ and $\tilde{E}_{21}$.

Thus, according to Eqs. (6.4a, 6.4b) we assume that the 2D anisotropy can be given by the following relations:

$$\tilde{S}_{12} = 2\tilde{\mu}_{12}\tilde{E}_{12}, \quad \tilde{S}_{21} = 2\tilde{\mu}_{21}\tilde{E}_{21} \tag{6.8}$$

We assume that fracture anisotropy is confined to the region in which the micro-fracturing processes proceed. Thus, we may consider two cases: the 2D uniform skew system anisotropy and the 2D fracture skew system anisotropy; the first is useful when considering the problem of wave propagation, while the other applies mainly to the part of continuum in which a fracturing develops from the micro-scale to the macro-processes.

For the microfracture processes we can write the related 2D wave equations for these asymmetric fields (cf., Teisseyre 2006, and corrected in Teisseyre 2008a, b); we write similar equations for $\tilde{E} \to \tilde{E}_{12}$, $V \to V_1$ and $\tilde{E} \to \tilde{E}_{21}$, $V \to V_2$; these relations will have the same form, in a skew anisotropic system, for the contravariant (parallel) and covariant (perpendicular) projections and anisotropic velocity, namely,

$$\tilde{E}\Big|_K^K - \frac{1}{V^2}\frac{\partial^2}{\partial t^2}\tilde{E} = 0, \; \tilde{E}\Big|_K^K \equiv \frac{\partial^2 \tilde{E}}{\partial X^K \partial X_K} = \frac{\partial^2 \tilde{E}}{g_{KS}\partial X^K \partial X^S} \text{ and } V^2 = \sum_{K,S} g_{KS} V^K V^S \tag{6.9}$$

where for the related type of solution, $\tilde{E} = E^0 \exp[\mathrm{i}(K_S X^S - \varpi t)]$, we should put $g^{KS} K_K K_S - \bar{\omega}^2 = 0$.

With these assumptions we obtain, for both spin (rotation) and twist (shear) motions, the two wave equations at the possible fractures planes $X^2 = 0$ and $X^1 = 0$ with the different velocities, $V_1^2 = \frac{2\mu_1}{\rho}$ and $V_2^2 = \frac{2\mu_2}{\rho}$ (assuming that anisotropy is related only to the rigidity modulus):

$$\tilde{E}_{12}\big|_K^K - \frac{1}{V_1^2}\frac{\partial^2}{\partial t^2}\tilde{E}_{12} = 0, \; \tilde{E}_{21}\big|_K^K - \frac{1}{V_2^2}\frac{\partial^2}{\partial t^2}\tilde{E}_{21} = 0 \tag{6.10}$$

We arrive at solutions with different waves in the skew coordinate system:

$$\tilde{E}_{12} = E^0_{12} \exp\left[\mathrm{i}\left(K'_S X^S - \varpi t\right)\right], \quad \tilde{E}_{21} = \tilde{E}^0_{21} \exp\left[\mathrm{i}\left(K''_S X^S - \varpi t\right)\right] \tag{6.11}$$

where we should put $g'^{KS} K'_K K'_S - \varpi^2 = 0$ and $g''^{KS} K''_K K''_S - \varpi^2 = 0$.

For our case, we obtain, for both spin (rotation) and twist (shear) motions, the two wave equations with different velocities, and we have arrived at solutions with different wave vectors, K', K'', in the skew coordinate system:

$$\tilde{E}_{12} = E^0_{12} \exp\left[\mathrm{i}\left(K'_S X^S - \varpi t\right)\right], \quad \tilde{E}_2 \equiv \tilde{E}^0_{21} = E^0_2 \exp\left[\mathrm{i}\left(K''_S X^S - \varpi t\right)\right] \tag{6.12}$$

Further on, we may consider the 2D fracture skew anisotropy; usually, we observe the occurrence of a privileged fracture plane and a system of weaker skew-oriented auxiliary fractures. The related anisotropy can be expressed again by the 2D skew coordinate system X^K ($K = \{1, 2\}$), with an angle φ between the axes X^1 and x_1 related to the rectangular system $\{x_1, x_2\}$, and angle ψ between the axes X^2 and x_2 (Fig. 6.2).

We may note that for a sequence of such systems with the variable angles φ between the consecutives skew axes we may arrive at the 2D curvilinear coordinates and related anisotropy; of course, we have started with the initial orthogonal system.

Now we can write the general formula valid for any direction in plane (X^1, X^2):

$$\tilde{E}\big|_K^K - \frac{1}{V^2}\frac{\partial^2}{\partial t^2}\tilde{E} = 0 \quad \text{for} \quad \tilde{E}\big|_K^K = \frac{\partial^2 \tilde{E}}{g_{KS}\partial X^K \partial X^S} \tag{6.13a}$$

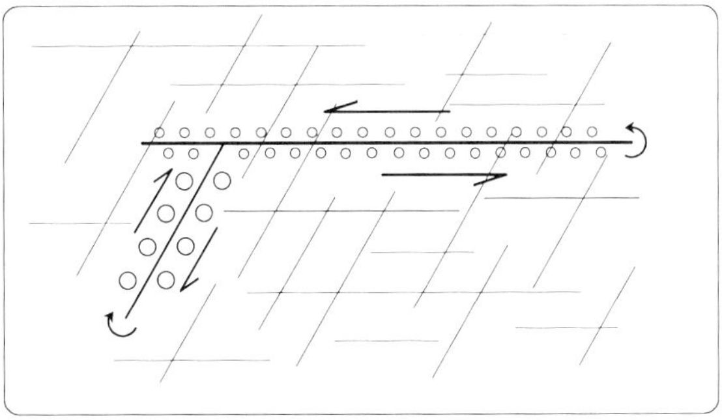

Fig. 6.2 Explanation to the 2D skew system anisotropy at fracture processes

and

$$V^2 = g_{11}V^1V^1 + g_{22}V^2V^2 + 2g_{12}V^1V^2 = V^1V^1\cos^2\alpha_1 + V^2V^2\cos^2\alpha_2 + 2V^1V^2\cos\alpha_1\cos\alpha_2$$
$$\mu = g_{11}\mu_1 + g_{22}\mu_2 + 2g_{12}\sqrt{\mu_1\mu_2} \tag{6.13b}$$

Further, we may write the related constitutive laws separately for the anti-symmetric and symmetric self-stresses and self-rotations (cf., Eq. 6.12):

$$\tilde{S}_{[ki]} = \sum_{m=1,2} \tilde{\mu}_{[ki]m}\tilde{E}_m, \quad \tilde{S}_{(ki)} = \sum_{m=1,2} \tilde{\mu}_{(ki)m}\tilde{E}_m, \quad \tilde{S}_{ki} = \tilde{S}_{[ki]} + \tilde{S}_{(ki)} \tag{6.14}$$

With these expressions we can enter into the respective equation of motion for the antisymmetric stresses (stress moment related balance equation) and then, following the procedure usually applied for anisotropy, we arrive at the eikonal equation. However, instead of this approach to anisotropy based on the eikonal equation, we may use a more direct geometrical approach as already considered (see: Teisseyre 1955) and the ray theory (see: Yajima and Nagahama 2006).

Thus, it seems convenient to express the 2D anisotropy, as related to the antisymmetric part of stresses (stress moment anisotropy) and strains, using the related skew coordinate system.

The fracture processes trace the pre-slip structure of media revealed in the formation of mylonite layers; the mylonite zones form an anisotropic system in which the difference in the grain radii causes the inertia-related anisotropy determining the fracture ability and friction properties. Figure 6.2 presents the skew system of the privileged slip planes having different ability for slip motions – as expressed by different radii of grains.

We assume that fracture anisotropy is confined to the region in which the microfracturing processes proceed; to this end, we redefine the constitutive law (6.8) and (6.13a, 6.14) to the constitutive relations in the 2D skew system:

$$\tilde{S}^{KS} = 2\tilde{\mu}^0_{KS}\tilde{E}^{KS}, \text{ indexes } KS = \{12, 21\} \tag{6.15a}$$

and moreover we include the effects of former microfractures as follows:

$$\tilde{S}^{KS} = 2\tilde{\mu}^0_{KS}\left[1 - p_1\exp\left(-q_1(X^1)^2\right) - p_2\exp\left(-q_2(X^2)^2\right)\right]\tilde{E}^{KS}, \quad \text{indexes } KS = \{12, 21\} \tag{6.15b}$$

In this definition we included the isotropic constitutive law and the functions describing the drops of the rotation rigidity occurring close to the planes $X^1 = 0$ and $X^2 = 0$. These drops may be characterized by the different rates (factors p_1, q_1 and p_2, q_2) as suggested and visualized by the different radii of grains in Fig. 6.2.

Concluding we recall our considerations on a relation between the antisymmetric stresses and rotations in a case of anisotropy: we have introduced geometrical approach to the problem of skew and fracture anisotropies. Of course, this is possible under the condition that our skew system is directly related to the considered anisotropy.

References

Jaroszewicz LR, Krajewski Z, Solarz L (2006) Absolute rotation measurement based on the Sagnac effect. In: Teisseyre R, Takeo M, Majewski E (eds) Earthquake source asymmetry, structural media and rotation effects. Springer, New York, pp 413–438

Kröner E (1981) Continuum theory of defect. In: Balian R, Kleman M, Poirer JP (eds) Les Houches, Session XXXV, 1980, Physics of Defects. North Holland, Amsterdam

Moriya T, Teisseyre R (2006) Design of rotation seismometer and non-linear behaviour of rotation components of earthquakes. In: Teisseyre R, Takeo M, Majewski E (eds) Earthquake source asymmetry, structural media and rotation effects. Springer, Berlin, pp 439–450

Schreiber KU, Stedman GE, Igel H, Flaws A (2006) Ring laser gyroscopes as rotation sensors for seismic wave studies. In: Teisseyre R, Takeo M, Majewski E (eds) Earthquake source asymmetry, structural media and rotation effects. Springer, Berlin, pp 377–389

Shimbo M (1975) A geometrical formulation of asymmetric features in plasticity. Bull Fac Eng (Hokkaido University) 77:155–159.

Takeo M (2006) Ground rotational motions recorded in near-source region of earthquakes. In: Teisseyre R, Takeo M, Majewski E (eds) Earthquake source asymmetry, structural media and rotation effects. Springer, Berlin, pp 157–167

Teisseyre R (1955) Optico-geometrical approximation for seismic waves in non-homogeneous media. Acta Geophys Pol 3:161–166

Teisseyre R (2005) Asymmetric continuum mechanics: deviations from elasticity and symmetry. Acta Geophys Polon 53:115–126

Teisseyre R (2006) Asymmetric continuum and anisotropy. Acta Geophys Pol 54(3):225–238

Teisseyre R (2008a) Introduction to asymmetric continuum: dislocations in solids and extreme phenomena in fluids. Acta Geophys 56:259–269

Teisseyre R (2008b) Asymmetric continuum: standard theory. In: R Teisseyre, H Nagahama and E Majewski (eds.) Physics of asymmetric continua : extreme and fracture processes, Springer, Berlin, pp 95–109

Teisseyre R, Górski M, Teisseyre KP (2006) Fracture-band geometry and rotation energy release. In: Teisseyre R, Takeo M, Majewski E (eds) Earthquake source asymmetry structural media and rotation effects. Springer, Berlin, pp 169–184

Wiszniowski J (2006) Rotation and twist motion recording—couple pendulum and rigid seismometers system. In: Teisseyre R, Takeo M, Majewski E (eds) Earthquake source asymmetry, structural media and rotation effects. Springer, Berlin, pp 451–468

Yajima T, Nagahama H (2006) Seismic ray theory for structural medium based on Kawaguchi and Finsler geometry. In: Teisseyre R, Takeo M, Majewski E (eds) Earthquake source asymmetry, structural media and rotation effects. Springer, Berlin, pp 329–336

Chapter 7
Induced Strains

7.1 Introduction

Any real continuum has some point and linear defects; our aim is to study a mechanism of reorganization of the applied stress load due to an influence of defects. Such an influence may be related both to the natural defects and, in the upper crust layers, also those related to the human activity; this is especially important for the mining regions and water dam areas. In general, the defects enhance an effect of stress load and related deformations. In the case an axial stress load is applied, such processes help to understand the formation of shear or rotational micro-fractures; in this way we may better understand the processes recognizing fragmentation and slip motions. Thus, we should account for the appearance of micro-fractures having a quite different character than those related to an external load; such effect may appear due to the role of internal defects existing as an integral element of any natural material. In this way, we may better explain the forwarding micro-destruction processes before any seismic event. We consider here a number of different problems related to the defect distribution.

7.2 Induced Strains

Any real continuum has some point and linear defects. Moreover, for some problems we should consider also an influence of the defects caused by the human activity; of course, such problems relate only to quite shallow layers. In this chapter we consider a mechanism of reorganization of the applied stress load due to any defect content, thus, we include here also some human activity.

The defects may enhance the effect of stress load and related deformations. In particular, the defect distributions reorganize an internal stress distribution, and may influence the relation between seismic risk and deformation level. Induced strains have an important role in mining regions and damp areas; using a continuum point of view we may directly consider an influence the internal defects, as

R. Teisseyre and M. Teisseyre-Jeleńska, *Asymmetric Continuum*,
GeoPlanet: Earth and Planetary Sciences, DOI: 10.1007/978-3-642-31860-3_7,

caused by the mining works, on possible fracture processes. The above-mentioned activity may cause an essential change of strain distribution and cause the fracture effects quite different from those related to the existing external load. In the case of an axial external load such a mechanism helps to understand a formation of shear or rotational micro-fractures; such processes lead to recognized fragmentation and slip motions.

Some material micro-destructions preceding the earthquakes or induced seismic events relate to very complicated nonlinear processes. The appearing micro-fractures may have a quite different character than that related to an external load; this is due to the role of internal defects existing as an integral element of any natural material. Thus, we may better understand the forwarding micro-destruction processes before any seismic event; the different defect densities may explain different effects caused by a given load.

Therefore, we should take into account the different effects than those expected due to the given boundary or initial conditions at action of the axial, deviatoric (shear) and antisymmetric (moment related rotation) strains. Thus, these effects are related to an unknown distribution of the internal defects and its changes due to human activity.

The Peach–Koehler forces acting on the defects may help us to define the induced strains, an effect related to the applied stress load in a body with internal defects. Defect density alters many bulk parameters of a body; in particular, it helps us to understand the formation of micro-fractures and fragmentations. Induced fields appear due to defects; in a continuum we may consider the defect densities. These induced fields can be explained, in an exact manner, when considering the Peach–Koehler forces exerted on defects (Peach and Koehler 1950); the forces exerted on dislocations can be presented as follows:

$$F_n = \varepsilon_{nsq} \sum_k S_{sk} b_k v_q \tag{7.1a}$$

where b_k is the dislocation slip vector and v_q is the dislocation line versor related to a defect line.

For the specific stress parts, $\bar{S}, \hat{S}, \breve{S}$ (axial, deviatorie shear and rotation stresses), we obtain:

$$\begin{aligned} \bar{F}_n &= \varepsilon_{nsq}\bar{S} \sum \delta_{sk} b_k v_q = \varepsilon_{nkq}\bar{S} b_k v_q, \quad \hat{F}_n = \varepsilon_{nsq} \sum_k \hat{S}_{(sk)} b_k v_q, \\ \breve{F}_n &\rightarrow \frac{1}{l} M_n = \varepsilon_{nsq} \sum_k \breve{S}_{[sk]} b_k v_q \end{aligned} \tag{7.1b}$$

The Peach–Koehler force, F_n, acts on defects; the Peach–Koehler relation presents the mutual interaction between the stresses and forces exerted on dislocations. These forces cause motions of defects to achieve some equilibrium positions; such a redistribution of defects may also lead to micro-fracture processes related to defect density:

$$\bar{F}_n = \varepsilon_{nsq}\bar{S}\sum \delta_{sk}\alpha_{qk} = \varepsilon_{nkq}\bar{S}\alpha_{qk}, \quad \hat{F}_n = \varepsilon_{nsq}\sum_k \hat{S}_{(sk)}\alpha_{qk},$$
$$\breve{F}_n = \frac{1}{l}M_n = \varepsilon_{nsq}\sum_k \breve{S}_{[sk]}\alpha_{qk} \tag{7.1c}$$

where on the right-hand side we have generalized this expression putting instead of the product, $b_k v_q$, the defect density: $v_q b_k \to \alpha_{qk}$.

The induced stresses, S_{np}^{IND}, can be defined with the help of a normal to defect plane, n_p:

$$F_n n_p = \varepsilon_{nsq}\sum_k b_k v_q n_p S_{sk}^{LOAD} \quad \to \quad S_{np}^{IND} = \varepsilon_{nsq}\sum_k \alpha_{qk} n_p S_{sk}^{LOAD}$$
$$S_{ik}^{IND} = 1/2\left(S_{(ik)}^{IND} + S_{[ik]}^{IND}\right) \tag{7.2}$$

where on the right-hand side we have generalized this expression putting instead of the product, $b_k v_q$, the defect density: $v_q b_k \to \alpha_{qk}$.

Note that the intensity of the induced stresses depends directly on defect density and applied load. Here, we will discuss the effects of induced stresses caused by a defect concentration. An extreme defect concentration may break a material continuity. As a consequence of such breaks, the strain compatibility relations (meaning that the strains could be presented by the displacement derivatives, cf., Chap. 2) could be broken.

The total stress field inside a body is given as a sum of the applied and induced stresses:

$$S_{np} = S_{np}^{LOAD} + S_{np}^{IND} \tag{7.3}$$

Here, the total and the induced strains may not depend on any displacements, because the induced part of these stresses is related to more complicated defect fields. We may remember that the compatibility condition (see Chap. 2 Eq. 2.2b) assure that strains or deformations can be presented by the displacement derivatives and that the displacements are basically taken as fields existing only in the mathematical sense. However, the induced stresses may become more complicated.

Now we may consider some specific examples; from Eq. (7.2) we obtain the following equations:

$$S_{1p}^{IND} = \left(\alpha_{31}S_{21}^{LOAD} - \alpha_{21}S_{31}^{LOAD} + \alpha_{33}S_{23}^{LOAD} - \alpha_{22}S_{32}^{LOAD} + \alpha_{32}S_{22}^{LOAD} - \alpha_{23}S_{33}^{LOAD}\right)n_p$$
$$S_{2p}^{IND} = \left(\alpha_{12}S_{32}^{LOAD} - \alpha_{32}S_{12}^{LOAD} + \alpha_{11}S_{31}^{LOAD} - \alpha_{33}S_{13}^{LOAD} + \alpha_{13}S_{33}^{LOAD} - \alpha_{31}S_{11}^{LOAD}\right)n_p$$
$$S_{33}^{IND} = \left(\alpha_{23}S_{13}^{LOAD} - \alpha_{13}S_{23}^{LOAD} + \alpha_{22}S_{12}^{LOAD} - \alpha_{11}S_{21}^{LOAD} + \alpha_{21}S_{11}^{LOAD} - \alpha_{12}S_{22}^{LOAD}\right)n_p \tag{7.4a}$$

Here the order of indexes is fixed; we cannot split these expressions into symmetric and antisymmetric parts; to better explain it, we may write the following two expressions containing quite different terms:

$$S_{12}^{IND} = \left(\alpha_{31}n_2S_{21}^{LOAD} - \alpha_{21}n_2S_{31}^{LOAD}\right) + \left(\alpha_{33}n_2S_{23}^{LOAD} - \alpha_{22}n_2S_{32}^{LOAD}\right) + \left(\alpha_{32}n_2S_{22}^{LOAD} - \alpha_{23}n_2S_{33}^{LOAD}\right)$$
$$S_{21}^{IND} = \left(\alpha_{12}n_1S_{32}^{LOAD} - \alpha_{32}n_1S_{12}^{LOAD}\right) + \left(\alpha_{11}n_1S_{31}^{LOAD} - \alpha_{33}n_1S_{13}^{LOAD}\right) + \left(\alpha_{13}n_1S_{33}^{LOAD} - \alpha_{31}n_1S_{11}^{LOAD}\right) \tag{7.4b}$$

Some realistic estimations indicate that a geological material can depend on two-level structures with bigger and the smaller grains, and the defects thus formed can depend on the grain sizes of a considered material structure. Then, we can expect that the appearing breaks would be related to a space between the greater grains (double structures of grains is meet frequently in geological materials). As presented in Eq. (7.3) the induced defect concentration leads to the total strains which may reach an effective material strength.

As examples, we consider two similar problems related to a rectangular corridor and horizontal tunnel in rocks; these man-made macro-defects would progressively create and activate the defects (dislocations and disclinations) leading to formation of arrays and induced strains. Near the inner boundary of such macro-defects, corridor or tunnel, there may be appear opposite-sense dislocation arrays related to the greater grain structure. Thus, close to the corridor or tunnel's outer boundary there can slowly be formed defect arrays and related growth of total strains.. The intensity of strains near the locked dislocations (top of arrays), nS, might slowly increase (due to an increase of the number of dislocations in an array, n).

The micro-defect arrays may form two opposite-sense defect segments, one located near a corridor/tunnel and the other farther from the tunnel; due to the pressure load we may expect an appearance of some defect arrays appearing in the vicinity of a corridor/tunnel perpendicularly to a given corridor (tunnel) sides (radial or vertical and horizontal arrays).

A time sequence of the induced stress increases should depend on the consecutive accumulations of defects (consecutive increases of the number of dislocations), as shown in Fig. 7.1.

Due to an increase of the defect densities in time we may expect that the sequence of the strain amplitudes related to the opposite arrays as functions of the distances from corridor or tunnel will grow. The consecutive figures image a slow increase of the defect densities. In this figure we observe the critical top points (de facto: a line along a tunnel corridor), where the defects of these two opposite kinds are accumulated. Together with the applied strain field and with those accumulation in arrays we may expect the maxima of stress accumulated (critical point "C") and then may have material fractures at the critical level, S^C:

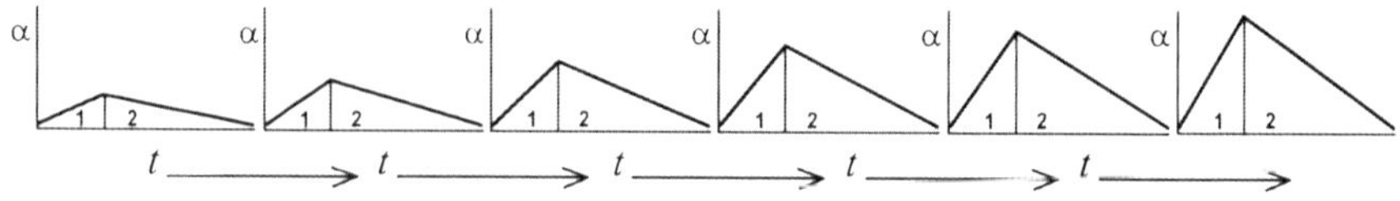

Fig. 7.1 The time sequence of the opposite-sense arrays (1,2), presenting increases of the array intensities in time; *parts 1*—the outer part of arrays, and *parts 2*—the defect increases in a deeper part

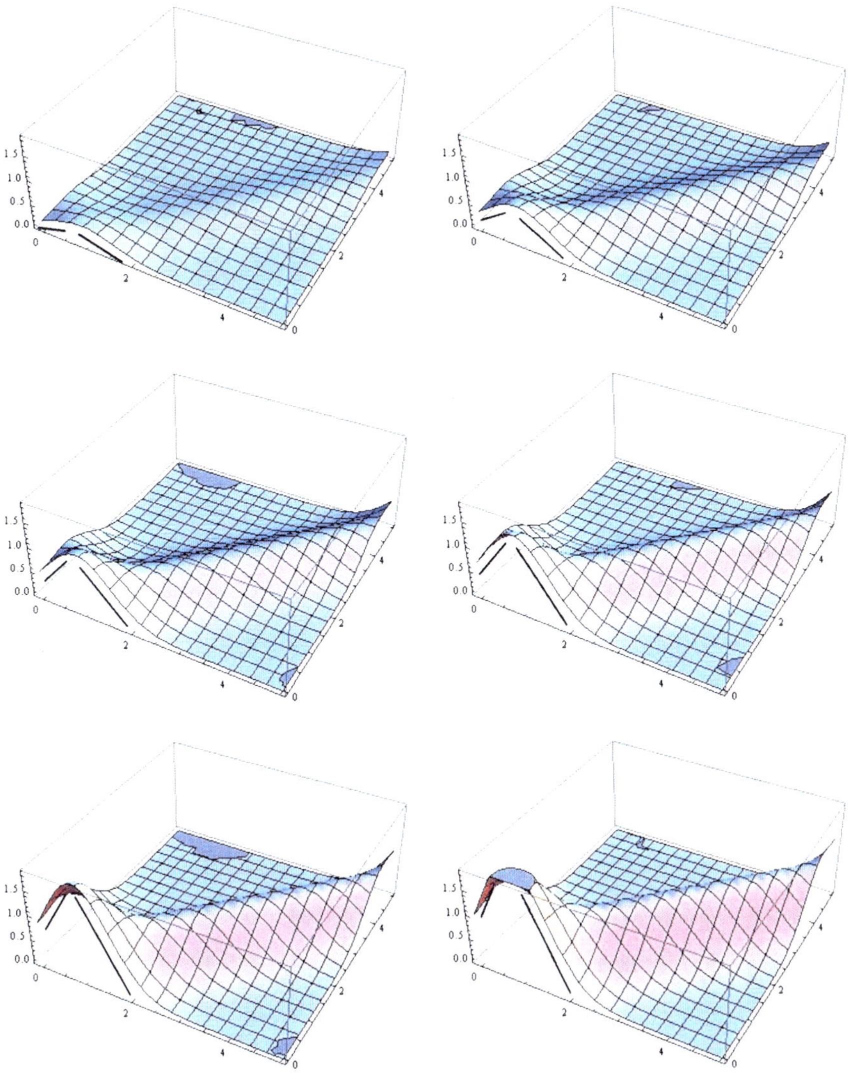

Fig. 7.2 Animation of the amplitudes of propagating waves for a sequence of six micro-fracture events (program—Mathematica 8); the consecutive parts of this figure present a slow increase of defect densities in two oppositely oriented arrays corresponding to Fig. 7.1, with a first array formed near a corridor/tunnel boundary and the opposite array more far from it

$$S^C = S^{LOAD} + \tilde{S}^C + \tilde{\tilde{S}}^C \tag{7.5}$$

where $\tilde{S}^C$ and $\tilde{\tilde{S}}^C$ present the stress accumulation in the first array (close to the corridor tunnel boundary), and in the second array (farther from the boundary).

When such micro-fractures appear, these micro-breaks give rise to the strain waves. In Fig. 7.2 we demonstrate an animation of the sequence of the possible wave events caused by such breaks.

The released process could appear near the "critical" point C (where the two opposite dislocation arrays meet); the related wave frequency spectrum will depend on the simultaneous fractures in the both grain structures.

Our further examples relate to the induced strains around a rectangular corridor; we may assume some defect distribution in the rock material around a corridor. Considering the defect densities, α_{xx} or α_{xy} or α_{xz} and α_{yx} or α_{yy} or α_{yz} in the two opposite directions, we may assume that these defects appear at a constant external pressure:

$$p = -\left(S_{xx}^{LOAD} + S_{yy}^{LOAD} + S_{zz}^{LOAD}\right)/3 = \text{constant} \tag{7.6}$$

The induced stresses, $S_{ik}^{IND} = 1/2\left(S_{(ik)}^{IND} + S_{[ik]}^{IND}\right)$, having normal along the line x ($x = x_1$, $y = x_2$, $z = x_3$), will be given, according to Eq. (7.4a), as follows:

$$\begin{aligned} S_{11}^{IND} &= (\alpha_{32} - \alpha_{23})S^{LOAD}n_1 \approx \alpha_{32}S^{LOAD}n_1 \\ S_{21}^{IND} &= (\alpha_{13} - \alpha_{31})S^{LOAD}n_1 \approx \sim -\alpha_{31}S^{LOAD}n_1 \\ S_{31}^{IND} &= (\alpha_{21} - \alpha_{12})S^{LOAD}n_1 \end{aligned} \tag{7.7a}$$

Here and further on we neglect the defects with the Burgers vector along the z-axis tunnel-wise.

The induced stresses having the normal along the line y will be given as:

$$\begin{aligned} S_{12}^{IND} &= (\alpha_{32} - \alpha_{23})S^{LOAD}n_2 \approx \alpha_{32}S^{LOAD}n_2 \\ S_{22}^{IND} &= (\alpha_{13} - \alpha_{31})S^{LOAD}n_2 \approx -\alpha_{31}S^{LOAD}n_2 \\ S_{32}^{IND} &= (\alpha_{21} - \alpha_{12})S^{LOAD}n_2 \end{aligned} \tag{7.7b}$$

while the induced stresses having the normal along the z line will be given as:

$$\begin{aligned} S_{13}^{IND} &= (\alpha_{32} - \alpha_{23})S^{LOAD}n_3 \approx \alpha_{32}S^{LOAD}n_3 \\ S_{23}^{IND} &= (\alpha_{13} - \alpha_{31})S^{LOAD}n_3 \approx \sim -\alpha_{31}S^{LOAD}n_3 \\ S_{33}^{IND} &= (\alpha_{21} - \alpha_{12})S^{LOAD}n_3 \end{aligned} \tag{7.7c}$$

In Fig. 7.3 we present some defect densities around a corridor, which have the edge axis along the y and z axes.

We may also consider the induced strains around the tunnel; we will consider the cases for which we may always put $E_{zz} = \text{constant}$ and where we neglect the variations along the z-axis.

A defect distribution around a man-made horizontal tunnel may be considered at any angle φ of the concentric defect arrays, at the condition $E_{zz} = \text{constant}$ and when neglecting the variations along z-axis. We demand for the components E_{rr} and $E_{\varphi\varphi}$, the condition $V^P = \varpi/k_r$, of the P-wave velocity.

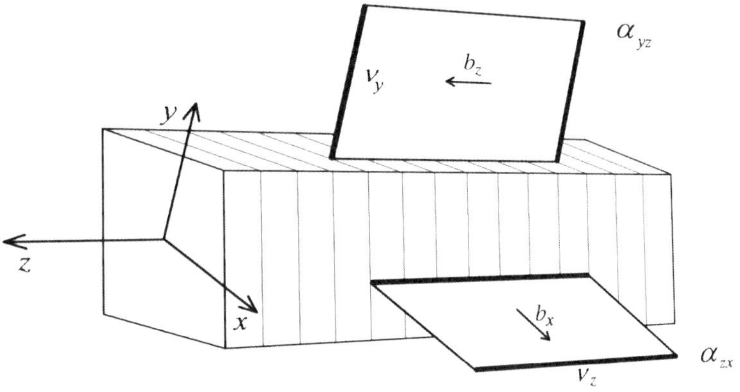

Fig. 7.3 Defect densities with the edge axis along the y direction, v_y, and along the z direction, v_z (these edges are marked by the *heavy lines*)

We consider the wave relations for these fields in the (r, φ, z) system; first, let us assume a micro-fracture at point $r = 0$ under the condition of the constant total strains, $E_{rr} + E_{\varphi\varphi} =$ constant. The possible wave oscillations of these components becomes given by the equations (cf., Chap. 2 and Eqs. 2.2 and 2.3):

$$\mu \sum_s \frac{\partial^2}{\partial x_s \partial x_s} E_{rr} - \rho \frac{\partial^2}{\partial t^2} E_{rr} = \frac{\partial F_r}{\partial r}, \quad \mu \sum_s \frac{\partial^2}{\partial x_s \partial x_s} E_{\varphi\varphi} - \rho \frac{\partial^2}{\partial t^2} E_{\varphi\varphi} = \frac{\partial F_\varphi}{r \partial \varphi} \quad (7.8a)$$

This leads to the Bessel solution for a radial, E_{rr}, and squeeze $E_{\varphi\varphi}$ components, as has been discussed in Chap. 2 and shown in Figs. 2.2 and 2.3.

Of special interest is the formation of the two opposite-direction arrays along the radial directions; the strain accumulation from oppositely directed arrays may lead, at the common top of arrays, to the critical point $C(r_C, \varphi)$, due to an accumulation of the induced stresses (cf. Fig. 7.1). However, near a tunnel boundary we should expect the strain falling to zero due to boundary conditions.

In Fig. 7.1 we have presented the opposite-directed arrays leading to micro-fracture; the energy releases due to annihilation process of the opposite-directed dislocations. Further, we present some figures visualizing some defect densities appearing around a tunnel (Fig. 7.4).

At the end we return to the problems of induced strains related to the defect densities of different kinds. We should remember that already in Chap. 2 we have considered the angular squeeze deformations, $E_{\varphi\varphi}$, in the horizontal plane as related to the cylindrical system, (r,φ,z), with the z-axis oriented in vertical direction (see Eqs. 2.15–2.21). The different defect densities play an important role in the formation induced strains. On the next diagrams gathered in Fig. 7.5 we illustrate such densities (remember that now the z-axis has the vertical orientation).

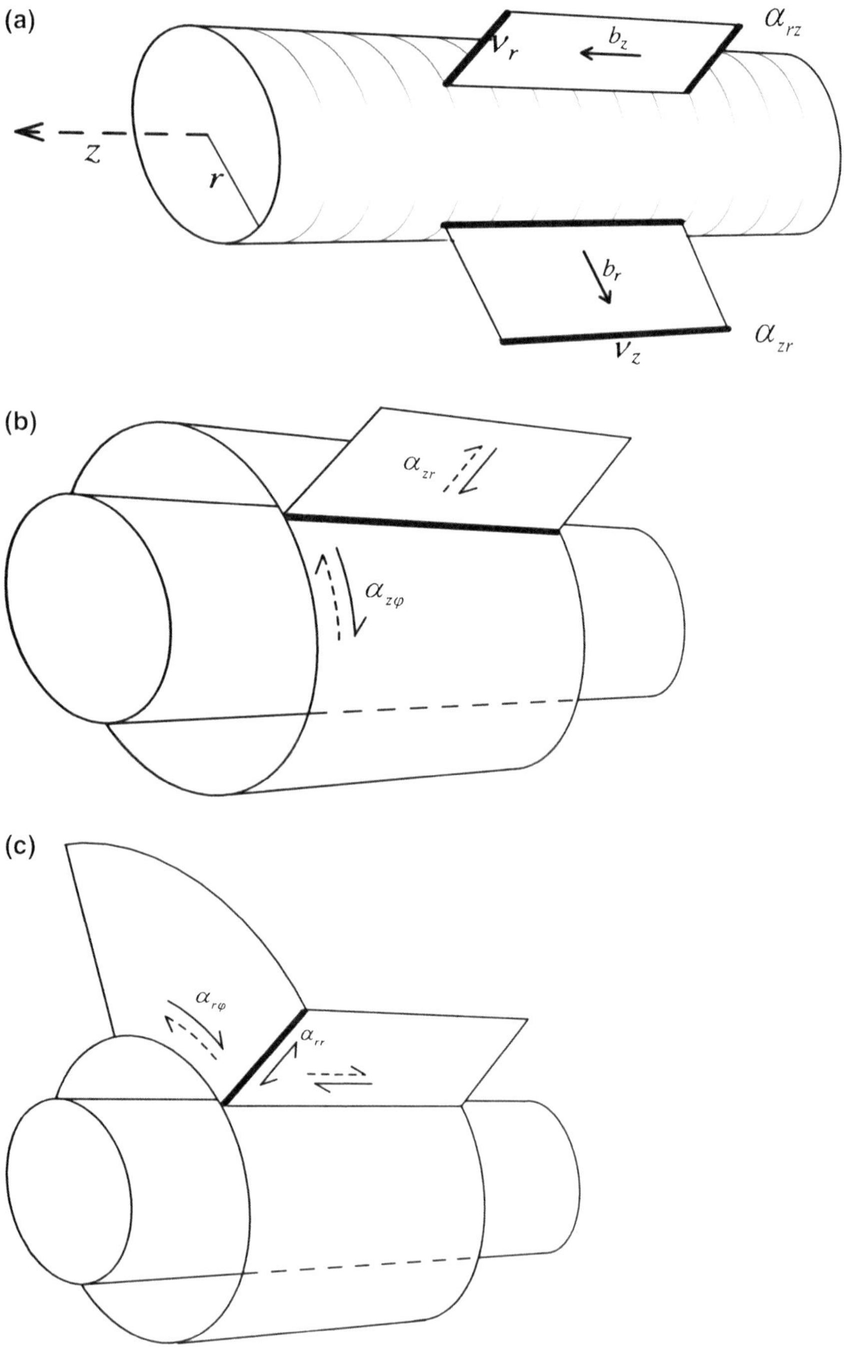

Fig. 7.4 **a** Radial defect distribution around a tunnel: α_{rz} and α_{zr}. **b** Defect densities α_{zr} (*upper part*) and $\alpha_{z\varphi}$ (*lower part*) around a tunnel. **c** Defect densities $\alpha_{r\varphi}$ and α_{rr}; there are also marked the densities α_{rz}, as may appear around a tunnel

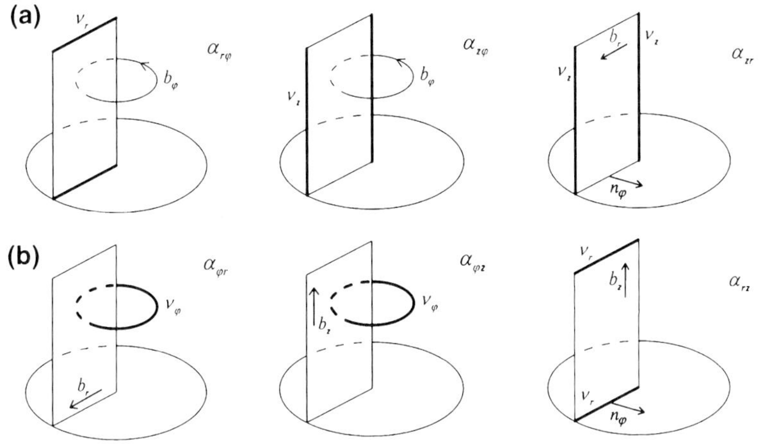

Fig. 7.5 Visualization of the defect densities: $\alpha_{r\varphi}$, $\alpha_{\varphi r}$, $\alpha_{z\varphi}$, $\alpha_{\varphi z}$, α_{zr}, α_{rz}

In the (r,φ,z) system we can write the definitions of the possible induced stresses (cf., Eqs. 7.2–7.4a):

$$
\begin{aligned}
S_{np}^{IND} &= \varepsilon_{nsq}\alpha_{qk}n_p S_{sk}^{LOAD}:\\
S_{rr}^{IND} &= \left(\alpha_{zr}S_{\varphi r}^{LOAD} - \alpha_{\varphi r}S_{zr}^{LOAD} + \alpha_{z\varphi}S_{\varphi\varphi}^{LOAD} - \alpha_{\varphi\varphi}S_{z\varphi}^{LOAD} + \alpha_{zz}S_{\varphi z}^{LOAD} - \alpha_{\varphi z}S_{zz}^{LOAD}\right)n_r\\
S_{r\varphi}^{IND} &= \left(\alpha_{zr}S_{\varphi r}^{LOAD} - \alpha_{\varphi r}S_{zr}^{LOAD} + \alpha_{z\varphi}S_{\varphi\varphi}^{LOAD} - \alpha_{\varphi\varphi}S_{z\varphi}^{LOAD} + \alpha_{zz}S_{\varphi z}^{LOAD} - \alpha_{\varphi z}S_{zz}^{LOAD}\right)n_\varphi\\
S_{rz}^{IND} &= \left(\alpha_{zr}S_{\varphi r}^{LOAD} - \alpha_{\varphi r}S_{zr}^{LOAD} + \alpha_{z\varphi}S_{\varphi\varphi}^{LOAD} - \alpha_{\varphi\varphi}S_{z\varphi}^{LOAD} + \alpha_{zz}S_{\varphi z}^{LOAD} - \alpha_{\varphi z}S_{zz}^{LOAD}\right)n_z\\
S_{\varphi r}^{IND} &= \left(\alpha_{rr}S_{zr}^{LOAD} - \alpha_{zz}S_{rz}^{LOAD} + \alpha_{r\varphi}S_{z\varphi}^{LOAD} - \alpha_{z\varphi}S_{r\varphi}^{LOAD} + \alpha_{rz}S_{zz}^{LOAD} - \alpha_{zr}S_{rr}^{LOAD}\right)n_r\\
S_{\varphi\varphi}^{IND} &= \left(\alpha_{rr}S_{zr}^{LOAD} - \alpha_{zr}S_{rr}^{LOAD} + \alpha_{r\varphi}S_{z\varphi}^{LOAD} - \alpha_{z\varphi}S_{r\varphi}^{LOAD} + \alpha_{rz}S_{zz}^{LOAD} - \alpha_{zz}S_{rz}^{LOAD}\right)n_\varphi\\
S_{\varphi z}^{IND} &= \left(\alpha_{rr}S_{zr}^{LOAD} - \alpha_{zr}S_{rr}^{LOAD} + \alpha_{r\varphi}S_{z\varphi}^{LOAD} - \alpha_{z\varphi}S_{r\varphi}^{LOAD} + \alpha_{rz}S_{zz}^{LOAD} - \alpha_{zz}S_{rz}^{LOAD}\right)n_z\\
S_{z\varphi}^{IND} &= \left(\alpha_{\varphi r}S_{rr}^{LOAD} - \alpha_{rr}S_{\varphi r}^{LOAD} + \alpha_{\varphi\varphi}S_{r\varphi}^{LOAD} - \alpha_{r\varphi}S_{\varphi\varphi}^{LOAD} + \alpha_{\varphi z}S_{rz}^{LOAD} - \alpha_{rz}S_{\varphi z}^{LOAD}\right)n_\varphi\\
S_{zr}^{IND} &= \left(\alpha_{\varphi r}S_{rr}^{LOAD} - \alpha_{rr}S_{\varphi r}^{LOAD} + \alpha_{\varphi\varphi}S_{r\varphi}^{LOAD} - \alpha_{r\varphi}S_{\varphi\varphi}^{LOAD} + \alpha_{\varphi z}S_{rz}^{LOAD} - \alpha_{rz}S_{\varphi z}^{LOAD}\right)n_r\\
S_{zz}^{IND} &= \left(\alpha_{\varphi r}S_{rr}^{LOAD} - \alpha_{rr}S_{\varphi r}^{LOAD} + \alpha_{\varphi\varphi}S_{r\varphi}^{LOAD} - \alpha_{r\varphi}S_{\varphi\varphi}^{LOAD} + \alpha_{\varphi z}S_{rz}^{LOAD} - \alpha_{rz}S_{\varphi z}^{LOAD}\right)n_z
\end{aligned}
\tag{7.9a}
$$

or in an abbreviated form:

$$
\begin{aligned}
S_{np}^{IND} &= \varepsilon_{nsq}\alpha_{qk}n_p S_{sk}^{LOAD} \; : \; n_n = (n_r, n_\varphi, n_z) \\
S_{rn}^{IND} &= \left(\alpha_{zr}S_{\varphi r}^{LOAD} - \alpha_{\varphi r}S_{zr}^{LOAD} + \alpha_{z\varphi}S_{\varphi\varphi}^{LOAD} - \alpha_{\varphi\varphi}S_{z\varphi}^{LOAD} + \alpha_{zz}S_{\varphi z}^{LOAD} - \alpha_{\varphi z}S_{zz}^{LOAD}\right)n_n \\
S_{\varphi n}^{IND} &= \left(\alpha_{rr}S_{zr}^{LOAD} - \alpha_{zz}S_{rz}^{LOAD} + \alpha_{r\varphi}S_{z\varphi}^{LOAD} - \alpha_{z\varphi}S_{r\varphi}^{LOAD} + \alpha_{rz}S_{zz}^{LOAD} - \alpha_{zr}S_{rr}^{LOAD}\right)n_n \\
S_{zn}^{IND} &= \left(\alpha_{\varphi r}S_{rr}^{LOAD} - \alpha_{rr}S_{\varphi r}^{LOAD} + \alpha_{\varphi\varphi}S_{r\varphi}^{LOAD} - \alpha_{r\varphi}S_{\varphi\varphi}^{LOAD} + \alpha_{\varphi z}S_{rz}^{LOAD} - \alpha_{rz}S_{\varphi z}^{LOAD}\right)n_n
\end{aligned}
\tag{7.9b}
$$

In particular, the related induced stresses acting on loop dislocations, with the normal n_φ, may be related to a contraction or expansion, $\alpha_{\varphi r}$, and to rotation around its axis, $\alpha_{\varphi\varphi}$, in its own plane, and also to the normal forces on $\alpha_{\varphi z}$:

$$
\begin{aligned}
S_{rz}^{IND} &= -\left(\alpha_{\varphi r}S_{zr}^{LOAD} + \alpha_{\varphi\varphi}S_{z\varphi}^{LOAD} + \alpha_{\varphi z}S_{zz}^{LOAD}\right)n_z \\
S_{zz}^{IND} &= \left(\alpha_{\varphi r}S_{rr}^{LOAD} + \alpha_{\varphi\varphi}S_{r\varphi}^{LOAD} + \alpha_{\varphi z}S_{rz}^{LOAD}\right)n_z
\end{aligned}
\tag{7.10}
$$

Further on we assume that the external load is related to pressure condition, which can be described by the axial stresses:

$$
S^{LOAD} = S_{rr}^{LOAD} = S_{\varphi\varphi}^{LOAD} = S_{zz}^{LOAD} = -p \tag{7.11}
$$

In particular, by virtue of Eq. (7.9b), the induced stresses having normal along an angular line, r, will be given as:

$$
\begin{aligned}
S_{rr}^{IND} &= (\alpha_{z\varphi} - \alpha_{\varphi z})S^{LOAD}n_r \approx \alpha_{z\varphi}S^{LOAD} \\
S_{\varphi r}^{IND} &= (\alpha_{rz} - \alpha_{zr})S^{LOAD}n_r \approx -\alpha_{zr}S^{LOAD} \\
S_{zr}^{IND} &= (\alpha_{\varphi r} - \alpha_{r\varphi})S^{LOAD}n_r
\end{aligned}
\tag{7.11a}
$$

where here and further on we neglect the defects with the Burgers vector along the z-axis.

The induced stresses having normal along an angular line φ will be given as:

$$
\begin{aligned}
S_{r\varphi}^{IND} &= (\alpha_{z\varphi} - \alpha_{\varphi z})S^{LOAD}n_\varphi \approx \alpha_{z\varphi}S^{LOAD}n_\varphi \\
S_{\varphi\varphi}^{IND} &= (\alpha_{rz} - \alpha_{zr})S^{LOAD}n_\varphi \approx -\alpha_{zr}S^{LOAD}n_\varphi \\
S_{z\varphi}^{IND} &= (\alpha_{\varphi r} - \alpha_{r\varphi})S^{LOAD}n_\varphi
\end{aligned}
\tag{7.11b}
$$

and the induced stresses having normal along an angular line, z, will be given as:

$$
\begin{aligned}
S_{rz}^{IND} &= (\alpha_{z\varphi} - \alpha_{\varphi z})S^{LOAD}n_z \approx \alpha_{z\varphi}S^{LOAD} \\
S_{\varphi z}^{IND} &= (-\alpha_{zr} + \alpha_{rz})S^{LOAD}n_z \approx -\alpha_{zr}S^{LOAD} \\
S_{zz}^{IND} &= (\alpha_{\varphi r} - \alpha_{r\varphi})S^{LOAD}n_z
\end{aligned}
\tag{7.11c}
$$

These induced stresses added to the applied stresses present the effective load acting in a considered region with the given defect densities.

In particular, in virtue of Eq. (7.9b), the induced stresses having the Bungers vector along angular line, r, will be given as:

$$\begin{aligned} S_{\varphi r}^{IND} &= -\alpha_{zr} S_{rr}^{LOAD} n_r, \quad S_{zr}^{IND} = \alpha_{\varphi r} S_{rr}^{LOAD} n_r \\ S_{\varphi\varphi}^{IND} &= -\alpha_{zr} S_{rr}^{LOAD} n_\varphi, \quad S_{z\varphi}^{IND} = \alpha_{\varphi r} S_{rr}^{LOAD} n_\varphi \\ S_{\varphi z}^{IND} &= -\alpha_{zr} S_{rr}^{LOAD} n_z, \quad S_{zz}^{IND} = \alpha_{\varphi r} S_{rr}^{LOAD} n_z \end{aligned} \tag{7.12a}$$

while the induced stresses having the Burgers vector along angular line, φ, will be given as:

$$\begin{aligned} S_{rr}^{IND} &= \alpha_{z\varphi} S^{LOAD} n_r, \quad S_{zr}^{IND} = -\alpha_{r\varphi} S^{LOAD} n_r \\ S_{r\varphi}^{IND} &= \alpha_{z\varphi} S^{LOAD} n_\varphi, \quad S_{z\varphi}^{IND} = -\alpha_{r\varphi} S^{LOAD} n_\varphi \\ S_{rz}^{IND} &= \alpha_{z\varphi} S^{LOAD} n_z, \quad S_{zz}^{IND} = -\alpha_{r\varphi} S^{LOAD} n_z \end{aligned} \tag{7.12b}$$

However, we may neglect the defect densities with the Burgers vector along the z-line, and therefore the following induced stresses are considered as close to zero:

$$\begin{aligned} S_{rr}^{IND} &= -\alpha_{\varphi z} S^{LOAD} n_r \approx 0, \quad S_{\varphi r}^{IND} = \alpha_{rz} S^{LOAD} n_r \approx 0 \\ S_{r\varphi}^{IND} &= -\alpha_{\varphi z} S^{LOAD} n_\varphi \approx 0, \quad S_{\varphi\varphi}^{IND} = \alpha_{rz} S^{LOAD} n_\varphi \approx 0 \\ S_{rz}^{IND} &= -\alpha_{\varphi z} S^{LOAD} n_z \approx 0, \quad S_{\varphi z}^{IND} = \alpha_{rz} S^{LOAD} n_z \approx 0 \end{aligned} \tag{7.12c}$$

For the angular squeeze (circumferential deformation), see Chap. 2, we obtain:

$$S_{\varphi\varphi}^{IND} = \left(\alpha_{rz} S_{zz}^{LOAD} - \alpha_{zr} S_{rr}^{LOAD}\right) n_\varphi \tag{7.13}$$

We have developed relations for induced stresses as appearing due to a defect content existing in any geological and other common materials. We considered the Peach–Koehler forces acting on dislocations. The induced stresses together with the applied stress cause motions of defects leading to some equilibrium, and subsequently producing fractures and finally a fracture process.

This mechanism explains the fragmentation processes appearing under a constant pressure; in such a case the crucial role is played by a much lower strength for shears than that for a pure compression.

The considered special cases related to the induced internal stresses may also be applied for the vertical tunnels, e.g., boreholes; in such a case we should include a dependence of the used fields on the vertical component z.

We may include the effects related to the loop defects expressed in the (r,φ,θ) coordinate system.

At the end we should keep in mind that the material micro-destructions preceding the earthquakes or induced seismic events relate to very complicated, surely nonlinear processes.

A character of such micro-fractures may be quite different from that related to the existing external load; this is due to a role of the internal defects existing as an integral element of any natural material. Therefore, to better understand the forwarding micro-destruction processes before any seismic event, we must account for the distribution of different defect densities.

Under shear load, $S_{ss}^{LOAD}=0$, we obtain quite different relations under the condition that the induced axial strain will disappear too:

$$\begin{aligned}
S_{np}^{IND} &= \varepsilon_{nsq}\alpha_{qk}n_p S_{sk}^{LOAD}\ , S_{ss}^{LOAD}=0\ ;\ \text{e.g.}:\\
S_{12}^{IND} &= \left(\alpha_{31}n_2S_{21}^{LOAD}-\alpha_{21}n_2S_{31}^{LOAD}\right)+\left(\alpha_{33}n_2S_{23}^{LOAD}-\alpha_{22}n_2S_{32}^{LOAD}\right)+\left(\alpha_{32}n_2S_{22}^{LOAD}-\alpha_{23}n_2S_{33}^{LOAD}\right)\\
S_{21}^{IND} &= \left(\alpha_{12}n_1S_{32}^{LOAD}-\alpha_{32}n_1S_{12}^{LOAD}\right)+\left(\alpha_{11}n_1S_{31}^{LOAD}-\alpha_{33}n_1S_{13}^{LOAD}\right)+\left(\alpha_{13}n_1S_{33}^{LOAD}-\alpha_{31}n_1S_{11}^{LOAD}\right)\\
S_{11}^{IND} &= \left(\alpha_{31}S_{21}^{LOAD}-\alpha_{21}S_{31}^{LOAD}+\alpha_{33}S_{23}^{LOAD}-\alpha_{22}S_{32}^{LOAD}+\alpha_{32}S_{22}^{LOAD}-\alpha_{23}S_{33}^{LOAD}\right)n_1\\
S_{23}^{IND} &= \left(\alpha_{12}n_3S_{32}^{LOAD}-\alpha_{32}n_3S_{12}^{LOAD}\right)+\left(\alpha_{11}n_3S_{31}^{LOAD}-\alpha_{33}n_3S_{13}^{LOAD}\right)+\left(\alpha_{13}n_3S_{33}^{LOAD}-\alpha_{31}n_3S_{11}^{LOAD}\right)\\
S_{32}^{IND} &= \left(\alpha_{23}n_2S_{13}^{LOAD}-\alpha_{13}n_2S_{23}^{LOAD}\right)+\left(\alpha_{22}n_2S_{12}^{LOAD}-\alpha_{11}n_2S_{21}^{LOAD}\right)+\left(\alpha_{21}n_2S_{11}^{LOAD}-\alpha_{12}n_2S_{22}^{LOAD}\right)\\
S_{22}^{IND} &= \left(\alpha_{12}S_{32}^{LOAD}-\alpha_{32}S_{12}^{LOAD}+\alpha_{11}S_{31}^{LOAD}-\alpha_{33}S_{13}^{LOAD}+\alpha_{13}S_{33}^{LOAD}-\alpha_{31}S_{11}^{LOAD}\right)n_2\\
S_{31}^{IND} &= \left(\alpha_{23}n_1S_{13}^{LOAD}-\alpha_{13}n_1S_{23}^{LOAD}\right)+\left(\alpha_{22}n_1S_{12}^{LOAD}-\alpha_{11}n_1S_{21}^{LOAD}\right)+\left(\alpha_{21}n_1S_{11}^{LOAD}-\alpha_{12}n_1S_{22}^{LOAD}\right)\\
S_{13}^{IND} &= \left(\alpha_{31}n_3S_{21}^{LOAD}-\alpha_{21}n_3S_{31}^{LOAD}\right)+\left(\alpha_{33}n_3S_{23}^{LOAD}-\alpha_{22}n_3S_{32}^{LOAD}\right)+\left(\alpha_{32}n_3S_{22}^{LOAD}-\alpha_{23}n_3S_{33}^{LOAD}\right)\\
S_{33}^{IND} &= \left(\alpha_{23}S_{13}^{LOAD}-\alpha_{13}S_{23}^{LOAD}+\alpha_{22}S_{12}^{LOAD}-\alpha_{11}S_{21}^{LOAD}+\alpha_{21}S_{11}^{LOAD}-\alpha_{12}S_{22}^{LOAD}\right)n_3
\end{aligned} \tag{7.14}$$

while for the (r,φ,z) system we have at $S_{ss}^{LOAD}=0$:

$$\begin{aligned}
S_{np}^{IND} &= \varepsilon_{nsq}\alpha_{qk}n_pS_{sk}^{LOAD}\ :\\
S_{rr}^{IND} &= \left(\alpha_{zr}S_{\varphi r}^{LOAD}-\alpha_{\varphi r}S_{zr}^{LOAD}-\alpha_{\varphi\varphi}S_{z\varphi}^{LOAD}+\alpha_{zz}S_{\varphi z}^{LOAD}\right)n_r\\
S_{r\varphi}^{IND} &= \left(\alpha_{zr}S_{\varphi r}^{LOAD}-\alpha_{\varphi r}S_{zr}^{LOAD}-\alpha_{\varphi\varphi}S_{z\varphi}^{LOAD}+\alpha_{zz}S_{\varphi z}^{LOAD}\right)n_\varphi\\
S_{rz}^{IND} &= \left(\alpha_{zr}S_{\varphi r}^{LOAD}-\alpha_{\varphi r}S_{zr}^{LOAD}-\alpha_{\varphi\varphi}S_{z\varphi}^{LOAD}+\alpha_{zz}S_{\varphi z}^{LOAD}\right)n_z\\
S_{\varphi r}^{IND} &= \left(\alpha_{rr}S_{zr}^{LOAD}-\alpha_{zz}S_{rz}^{LOAD}+\alpha_{r\varphi}S_{z\varphi}^{LOAD}-\alpha_{z\varphi}S_{r\varphi}^{LOAD}\right)n_r\\
S_{\varphi\varphi}^{IND} &= \left(\alpha_{rr}S_{zr}^{LOAD}+\alpha_{r\varphi}S_{z\varphi}^{LOAD}-\alpha_{z\varphi}S_{r\varphi}^{LOAD}-\alpha_{zz}S_{rz}^{LOAD}\right)n_\varphi\\
S_{\varphi z}^{IND} &= \left(\alpha_{rr}S_{zr}^{LOAD}+\alpha_{r\varphi}S_{z\varphi}^{LOAD}-\alpha_{z\varphi}S_{r\varphi}^{LOAD}-\alpha_{zz}S_{rz}^{LOAD}\right)n_z\\
S_{z\varphi}^{IND} &= \left(-\alpha_{rr}S_{\varphi r}^{LOAD}+\alpha_{\varphi\varphi}S_{r\varphi}^{LOAD}+\alpha_{\varphi z}S_{rz}^{LOAD}-\alpha_{rz}S_{\varphi z}^{LOAD}\right)n_\varphi\\
S_{zr}^{IND} &= \left(-\alpha_{rr}S_{\varphi r}^{LOAD}+\alpha_{\varphi\varphi}S_{r\varphi}^{LOAD}+\alpha_{\varphi z}S_{rz}^{LOAD}-\alpha_{rz}S_{\varphi z}^{LOAD}\right)n_r\\
S_{zz}^{IND} &= \left(-\alpha_{rr}S_{\varphi r}^{LOAD}+\alpha_{\varphi\varphi}S_{r\varphi}^{LOAD}+\alpha_{\varphi z}S_{rz}^{LOAD}-\alpha_{rz}S_{\varphi z}^{LOAD}\right)n_z
\end{aligned} \tag{7.15a}$$

We may notice that the induced stresses related to the load having the Burgers vector along the angular line ϕ will be given as:

$$S_{rr}^{IND} = \left(-\alpha_{\varphi\varphi} S_{z\varphi}^{LOAD}\right) n_r, S_{r\varphi}^{IND} = \left(-\alpha_{\varphi\varphi} S_{z\varphi}^{LOAD}\right) n_\varphi, S_{rz}^{IND} = \left(-\alpha_{\varphi\varphi} S_{z\varphi}^{LOAD}\right) n_z$$
$$S_{\varphi r}^{IND} = \left(\alpha_{r\varphi} S_{z\varphi}^{LOAD} - \alpha_{z\varphi} S_{r\varphi}^{LOAD}\right) n_r, S_{\varphi\varphi}^{IND} = \left(\alpha_{r\varphi} S_{z\varphi}^{LOAD} - \alpha_{z\varphi} S_{r\varphi}^{LOAD}\right) n_\varphi, S_{\varphi z}^{IND} = \left(\alpha_{r\varphi} S_{z\varphi}^{LOAD} - \alpha_{z\varphi} S_{r\varphi}^{LOAD}\right) n_z$$
$$S_{z\varphi}^{IND} = \left(\alpha_{\varphi\varphi} S_{r\varphi}^{LOAD}\right) n_\varphi, S_{zr}^{IND} = \left(\alpha_{\varphi\varphi} S_{r\varphi}^{LOAD}\right) n_r, S_{zz}^{IND} = \left(\alpha_{\varphi\varphi} S_{r\varphi}^{LOAD}\right) n_z \tag{7.15b}$$

This is an interesting result which may be compared to the considerations on the $E_{\varphi\varphi}$ strains in Chap. 2. The last relations explain the introduced Peach–Koehler forces of new type acting on a ring (rotational) dislocation in its own plane: contraction or expansion and rotation around its axis.

An important influence on a source process and the premonitory micro-events is exerted by the material defects, their distribution and mobility. The defect arrays lead to a concentration of stresses and their local reorganization. In this book we consider the induced stresses and strains related to the content of defects and to its modification and redistribution.

The defect densities are closely related to the induced stresses; strains and defect densities distributed in the arrays may grow successively, leading to fracture; a single fracture episode may be defined as a mutual annihilation process of the defects, e.g., annihilation of the opposite-direction dislocations, or annihilation of the opposite ends of two cracks (joining of two cracks), cf., Droste and Teisseyre (1959). As we have presented, the induced stresses and strains appear due to defect content in a continuum; we should remember the Peach–Koehler forces exerted on defects (Peach and Koehler 1950), as appearing due to the applied stresses. However, we have generalized the original Peach–Koehler expression assuming, first, that the stresses, S_{sk}, represent an asymmetric field and, second, that we can describe the defect content by a continuous field, $\alpha_{qk} = v_q b_k$.

Due to a defect influence, as we have already shown, a sum of the applied stresses and induced ones leads to the total stress field:

$$S_{np}^{T} = S_{np}^{Load} + S_{np}^{Ind} = S_{(np)}^{Load} + S_{[np]}^{Load} + S_{(np)}^{Ind} + S_{[np]}^{Ind} \tag{7.16}$$

Of course, the total stresses would present the reorganized stress system.

We may consider the following specific cases:

- Let us assume that the total axial loads, $S_{np} \equiv S_{np}^{T} = S_{np}^{LOAD} + S_{np}^{IND}$ are given as follows: $S_{11} = S_{22} = S_{33} = \bar{S}$, $(S_{rr} = S_{\varphi\varphi} = S_{zz} = \bar{S})$ and for $S_{12} = S_{21} = 0$ $(S_{r\varphi} = S_{\varphi r} = 0)$ we obtain for the induced parts:

$$S_{kp} = S_{kp}^{Load} + S_{kp}^{Ind} = S_{kp}^{Load} - 3\varepsilon_{kns}\alpha_{ns}\bar{S} n_p;$$
$$S_{(kp)} = S_{(kp)}^{Load} + S_{(kp)}^{Ind} = -\frac{3}{2}\bar{S}\left(\varepsilon_{kns}\alpha_{ns} n_p + \varepsilon_{pns}\alpha_{ns} n_k\right), \tag{7.17}$$
$$S_{[kp]} = S_{[kp]}^{Ind} = -\frac{3}{2}\bar{S}\left(\varepsilon_{kns}\alpha_{ns} n_p - \varepsilon_{pns}\alpha_{ns} n_k\right)$$

for the loads, $S_{11}, S_{22}, (S_{rr}, S_{\varphi\varphi}), S_{33}{=}S_{zz}{=}\bar{\bar{S}}$, and $S_{12}{=}S_{21}{=}\ 0$, $(S_{r\varphi} = S_{\varphi r} = 0)$, we will obtain the following total stresses, $S^T_{np} = S_{np} + S^{Ind}_{np}$:

$$\begin{aligned} S_{1p} &= S^{Load}_{1p} + 3(\alpha_{32}S_{22} - \alpha_{23}\bar{S})n_p \\ S_{2p} &= S^{Load}_{2p} + 3(\alpha_{13}\bar{S} - \alpha_{31}S_{11})n_p \\ S_{3p} &= S^{Load}_{3p} + 3(\alpha_{21}S_{11} - \alpha_{12}S_{22})n_p \end{aligned} \tag{7.18a}$$

and

$$\begin{aligned} S_{rp} &= S^{Load}_{rp} + 3\left(\alpha_{z\varphi}S_{\varphi\varphi} - \alpha_{\varphi z}\bar{S}\right)n_p; \quad \text{for } p{=}\{r, \varphi, z\} \\ S_{\varphi p} &= S^{Load}_{\varphi p} + 3(\alpha_{rz}\bar{S} - \alpha_{zr}S_{rr})n_p; \quad \text{for } p{=}\{\varphi, z, r\} \\ S_{zp} &= S^{Load}_{zp} + 3\left(\alpha_{\varphi r}S_{rr} - \alpha_{r\varphi}S_{\varphi\varphi}\right)n_p; \quad \text{for } p{=}\{z, r, \varphi\} \end{aligned} \tag{7.18b}$$

We find that for the total axial loads, even beneath the corresponding strength value, the shear and fragmentation fracture processes can appear due to an influence of the defect co-action; however, in this case the obtained moment tensor solutions will not fit to the applied axial load.

In the above relations we have assumed a that the constant axial load before an event may exhibit some stress drop after it.

We conclude that an influence of the defect fields will effectively accelerate also the fracture processes for shear and angular moment loads.

However, as mentioned before, the emitted strain fields due to some event, or micro-event, and due to defects could be quite accidental; some part of the observed moment tensor solution (fault plane solutions) could deviate from the applied shear or moment loads in a different manner at different recording sites; probably it would be possible to select true solutions when using data from several recording sites.

We should keep in mind that for the shear and rotation processes, $\hat{E}$ and $\breve{E}$, we might observe the strain drops and the related waves appearing with the $\pi/2$ phase shift between these fields.

Let us return, again, to the action of pure confining stresses; when the load $\bar{S} = \frac{1}{3}\sum_k S_{kk}$ increases slowly, we may observe the appearance of a greater and greater number of defects and related induced stresses; the material properties undergo changes. To explain more clearly the role of defect content on formation of induced stresses under confining load we may use some simplification by imagining two artificial states: first, "purely elastic," related to increasing stresses and strains in a domain up to the upper limit of the "pure elasticity", and another, related to higher stresses and strains up to a fracture level at which an increase of confining load causes the defect increase (including reorganization of defect arrays).

This approach permits to better understand the role of defects in a progressive approach to fracture (or micro-fracture) under confining load; in real processes these two artificial states are combined with decreasing role of the first "elastic" state and increasing significance of a counterpart of defects in a way to fracture. The defect influence increases slowly at low stress load and quite rapidly in a final stage, close to fracture processes (we might also include the changes of the material properties and appearance of some material anisotropy).

In a final fracture process (under confining load) the shear and rotation stress release should be expected when these stresses will approach the respective strength level. This state can be achieved due to an increase of applied load and quite quick rise of a role of the defects. Thus, we repeat that, when approaching the fracture process, we will arrive at the shear and rotation fracture processes. The pressure strength is very elevated; therefore, due an increase of defect content and induces stresses, we arrive at the shear or/and fragmentation (concentric rotations) fracture processes:

$$S_{np}^{Ind} = \hat{S}_{(np)}^{Ind} + \overset{Ind}{\bar{S}}_{[np]} = -3\varepsilon_{nms}\alpha_{ms}\bar{S}n_p \approx \hat{S}_{(np)}^{Strength} + \overset{Strength}{\bar{S}}_{[np]} \quad ; \quad n \neq p \tag{7.19}$$

These fracture events relate to shear and fragmentation fractures and the wave radiation relates to the $\hat{E}_{(np)}^{Ind}$ and $\overset{Ind}{\bar{E}}_{[np]}$ fields, which means, they do not relate to the applied axial load.

Thus, we should be aware of the fact that the observed fault plane solutions will not relate directly to the axial load condition, but rather to the shear and fragmentation processes of different orientations (Fig. 7.6). For the processes under confining load, this statement is of great importance.

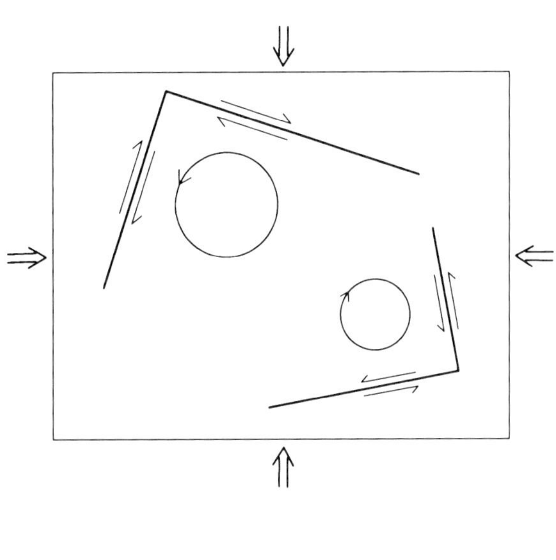

Fig. 7.6 Shear and fragmentations of different orientations due to the confining load

In the (r,θ,φ) system we obtain from Eq. (7.14) :

$$\begin{aligned}
S_{rp}^{T} &= S_{rp} + 3\left(\alpha_{\varphi r}S_{\theta r} - \alpha_{\theta r}S_{\varphi r} + \alpha_{\varphi\varphi}S_{\theta\varphi} - \alpha_{\theta\theta}S_{\varphi\theta} + \alpha_{\varphi\theta}S_{\theta\theta} - \alpha_{\theta\varphi}S_{\varphi\varphi}\right)n_p \\
S_{\theta p}^{T} &= S_{\theta p} + 3\left(\alpha_{r\theta}S_{\varphi\theta} - \alpha_{\varphi\theta}S_{r\theta} + \alpha_{rr}S_{\varphi r} - \alpha_{\varphi\varphi}S_{r\varphi} + \alpha_{r\varphi}S_{\varphi\varphi} - \alpha_{\varphi r}S_{rr}\right)n_p \\
S_{\varphi p}^{T} &= S_{\varphi p} + 3\left(\alpha_{\theta\varphi}S_{r\varphi} - \alpha_{r\varphi}S_{\theta\varphi} + \alpha_{\theta\theta}S_{r\theta} - \alpha_{rr}S_{\theta r} + \alpha_{\theta r}S_{rr} - \alpha_{r\theta}S_{\theta\theta}\right)n_p
\end{aligned} \tag{7.20a}$$

and for the shear and rotation stresses separately:

$$\begin{aligned}
\begin{Bmatrix} S_{(r\theta)}^{T} \\ S_{[r\theta]}^{T} \end{Bmatrix} &= \begin{Bmatrix} S_{(r\theta)} \\ S_{[r\theta]} \end{Bmatrix} + \frac{3}{2}\left(\alpha_{\varphi r}S_{\theta r} - \alpha S_{\varphi r} + \alpha_{\varphi\varphi}S_{\theta\varphi} - \alpha_{\theta\theta}S_{\varphi\theta} + \alpha_{\varphi\theta}S_{\theta\theta} - \alpha_{\theta\varphi}S_{\varphi\varphi}\right)n_\theta \\
&\quad \pm \frac{3}{2}\left(\alpha_{r\theta}S_{\varphi\theta} - \alpha_{\varphi\theta}S_{r\theta} + \alpha_{rr}S_{\varphi r} - \alpha_{\varphi\varphi}S_{r\varphi} + \alpha_{r\varphi}S_{\varphi\varphi} - \alpha_{\varphi r}S_{rr}\right)n_r \\
\begin{Bmatrix} S_{(r\varphi)}^{T} \\ S_{[r\varphi]}^{T} \end{Bmatrix} &= \begin{Bmatrix} S_{(r\varphi)} \\ S_{[r\varphi]} \end{Bmatrix} + \frac{3}{2}\left(\alpha_{\varphi r}S_{\theta r} - \alpha_{\theta r}S_{\varphi r} + \alpha_{\varphi\varphi}S_{\theta\varphi} - \alpha_{\theta\theta}S_{\varphi\theta} + \alpha_{\varphi\theta}S_{\theta\theta} - \alpha_{\theta\varphi}S_{\varphi\varphi}\right)n_\varphi \\
&\quad \pm \frac{3}{2}\left(\alpha_{\theta\varphi}S_{r\varphi} - \alpha_{r\varphi}S_{\theta\varphi} + \alpha_{\theta\theta}S_{r\theta} - \alpha_{rr}S_{\theta r} + \alpha_{\theta r}S_{rr} - \alpha_{r\theta}S_{\theta\theta}\right)n_r \\
\begin{Bmatrix} S_{(\theta\varphi)}^{T} \\ S_{[\theta\varphi]}^{T} \end{Bmatrix} &= \begin{Bmatrix} S_{(\theta\varphi)} \\ S_{[\theta\varphi]} \end{Bmatrix} + \frac{3}{2}\left(\alpha_{r\theta}S_{\varphi\theta} - \alpha_{\varphi\theta}S_{r\theta} + \alpha_{rr}S_{\varphi r} - \alpha_{\varphi\varphi}S_{r\varphi} + \alpha_{r\varphi}S_{\varphi\varphi} - \alpha_{\varphi r}S_{rr}\right)n_\varphi \\
&\quad \pm \frac{3}{2}\left(\alpha_{\theta\varphi}S_{r\varphi} - \alpha_{r\varphi}S_{\theta\varphi} + \alpha_{\theta\theta}S_{r\theta} - \alpha_{rr}S_{\theta r} + \alpha_{\theta r}S_{rr} - \alpha_{r\theta}S_{\theta\theta}\right)n_\theta
\end{aligned} \tag{7.20b}$$

We should note a counterpart of the angular-squeeze stress component, $S_{\varphi\varphi}$, the $\alpha_{\varphi\varphi}$ defects associated with to the formation of the related breaks (Teisseyre 2011; see Chap. 2).

As a particular case, we will consider the confining load condition, $\bar{S} = S_{rr} = S_{\theta\theta} = S_{\varphi\varphi}$, that may lead to some fracture event at the origin of the used coordinate system:

$$\begin{aligned}
S_{rp}^{T} &= 3\bar{S}\left(\alpha_{\varphi\theta} - \alpha_{\theta\varphi}\right)n_p; \quad n_p = \left\{n_\theta, n_\varphi\right\} : \\
S_{\theta p}^{T} &= 3\bar{S}\left(\alpha_{r\varphi} - \alpha_{\varphi r}\right)n_p; \quad n_p = \left\{n_\varphi, n_r\right\} : \\
S_{\varphi p}^{T} &= 3\bar{S}(\alpha_{\theta r} - \alpha_{r\theta})n_p; \quad n_p = \left\{n_r, n_\theta\right\} :
\end{aligned} \tag{7.20c}$$

Hence, for the shear and rotation stresses we obtain (cf., Eq. 7.19):

$$\begin{aligned}
\begin{Bmatrix} S_{(r\theta)}^{T} \\ S_{[r\theta]}^{T} \end{Bmatrix} &= \frac{3}{2}\bar{S}\left\{\left(\alpha_{\varphi\theta} - \alpha_{\theta\varphi}\right)n_\theta \pm \left(\alpha_{r\varphi} - \alpha_{\varphi r}\right)n_r\right\} \\
\begin{Bmatrix} S_{(r\varphi)}^{T} \\ S_{[r\varphi]}^{T} \end{Bmatrix} &= \frac{3}{2}\bar{S}\left\{\left(\alpha_{\varphi\theta} - \alpha_{\theta\varphi}\right)n_\varphi \pm (\alpha_{\theta r} - \alpha_{r\theta})n_r\right\} \\
\begin{Bmatrix} S_{(\theta\varphi)}^{T} \\ S_{[\theta\varphi]}^{T} \end{Bmatrix} &= \frac{3}{2}\bar{S}\left\{\left(\alpha_{r\varphi} - \alpha_{\varphi r}\right)n_\varphi \pm (\alpha_{\theta r} - \alpha_{r\theta})n_\theta\right\}
\end{aligned} \tag{7.20d}$$

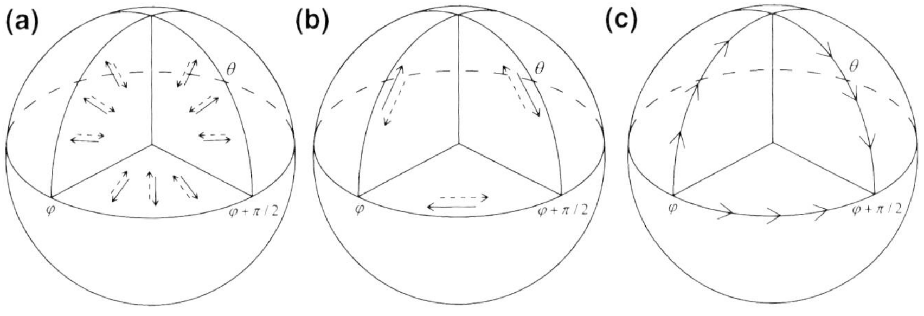

Fig. 7.7 Coaction processes. **a** Coaction of the induced shears. **b** Coaction of the rotation strain. **c** Coaction of the simple rotations

Considering the strains induced only due to the edge-type defect arrays ($\alpha_{\varphi r}$ and $\alpha_{\theta r}$—originally being formed on the perpendicular planes n_θ and n_φ) we obtain the strains on the perpendicular planes (here, total fields remain equal to the induced ones):

$$\left\{\begin{matrix} S^T_{(r\theta)} \\ S^T_{[r\theta]} \end{matrix}\right\} = \mp\frac{3}{2}\bar{S}\alpha_{\varphi r}n_r, \quad \left\{\begin{matrix} S^T_{(r\varphi)} \\ S^T_{[r\varphi]} \end{matrix}\right\} = \pm\frac{3}{2}\bar{S}\alpha_{\theta r}n_r,$$
$$\left\{\begin{matrix} S^T_{(\theta\varphi)} \\ S^T_{[\theta\varphi]} \end{matrix}\right\} = \frac{3}{2}\bar{S}\{-\alpha_{\varphi r}n_\varphi \pm \alpha_{\theta r}n_\theta\} \tag{7.20e}$$

These total stresses present only the induced parts appearing due to action of the confining load $\bar{S}$; such induced fields might be responsible for the shear and rotation fractures (fragmentations), as presented in Fig. 7.6.

The obtained stress system (7.20d) leads to some interesting correlations and mutual support in action of the induced strains and defect arrays. Thus, we can observe, as shown in Fig. 7.7 (parts a and b), that the defect arrays, $\alpha_{\theta r}$ (formed on the planes n_φ and $n_{\varphi+\pi/2}$), can support the formation of defect arrays, $\alpha_{\varphi r}$ (formed on the plane n_θ). Similarly, the strains $E^T_{(r\varphi)} = E^{Ind}_{(r\varphi)}$ (or/and $E^T_{[r\varphi]} = E^{Ind}_{[r\varphi]}$) on planes n_φ and $n_{\varphi+\pi/2}$, and strains $E^T_{(\theta\varphi)} = E^{Ind}_{(\theta\varphi)}$ (or/and $E^T_{[r\varphi]} = E^{Ind}_{[r\varphi]}$) on plane n_θ, form the mutually related deformation systems associated with the defect accumulation and shear strains (or/and rotation strains); as shown on this figure, the above-mentioned deformation systems are similar to an equivalence of rotations along the φ and θ paths (see Fig. 7.7 c).

Considering the induced seismicity we note that in some way we can include hereby the additional defects as caused by the mining works or due to water infiltrations.

Any induced field may appear due to the existence of suitable defects in a continuum; the induced energy will depend on the applied stress and strain for a homogeneous medium, where the stress considered shall be enhanced by a number of defects in a suitable array. This statement is taken from the formula presenting a linear stress increase at a tip of the defect array, e.g., related to the n dislocations.

References

Droste Z, Teisseyre R (1959) The mechanism of earthquakes according to dislocation theory. Sci Rep Tohoku Univ Ser Geophys 11(1):55–71

Peach M, Koehler JS (1950) The forces exerted on dislocations and the stress fields produced by them. Phys Rev 80:436–439

Teisseyre R (2011) Why rotation seismology: confrontation between classic and asymmetric theories. Bull Seismol Soc Am 101(4):1683–1691

Chapter 8
Thermodynamic Relations

8.1 Introduction

Keeping in mind that the main elements of the classic thermodynamics remain valid when consdering the Asymmetric Strain Theory, we will focus here only on some necessary corrections related to the rotational components appearing due to the presented new approach to the motions and deformations. In our considerations, important influences on the thermodynamic parameters in solids have been exerted by defects; therefore, we follow the cBΩ thermodynamical model (Varotsos and Alexopoulos, 1986), confirmed by a number of empirical relations and leading, among other things, to an important search of earthquake precursors. We define also the superlattice related to a distribution of the linear defects. The formation of a defect line leads to the negative contribution to the Gibbs free energy; therefore, we cannot reach a minimum for any number of the dislocations; however, we can omit this problem when introducing the dislocation superlattice. Moreover, we adopt the standard approach to the expression for released energies which contains, besides the symmetric stresses, also their antisymmetric parts; thus, the energy drops may include the rotational stress fields as well.

8.2 Thermodynamic Functions

The main elements of the classic thermodynamics remain valid when considering the Asymmetric Strain Theory; therefore, we will focus only on some necessary corrections and improvements appearing due to the presented new approach related to the motions and deformations. Here, we will not consider the known thermoelastic relations used for the displacements and temperature.

First let us specify the notation. The first law of thermodynamics relates to an increase of internal energy, U, of a body at a volume, V, and under applied stresses, S, strain changes, dE, and a heat received, dQ:

$$dU = dQ + (SdE)V \tag{8.1a}$$

R. Teisseyre and M. Teisseyre-Jeleńska, *Asymmetric Continuum*,
GeoPlanet: Earth and Planetary Sciences, DOI: 10.1007/978-3-642-31860-3_8,

where we have:

$$U = Q + \frac{1}{2}(SE)V;\ dU = dQ + \frac{1}{2}(SdE)V,\ dW = (SdE)V \tag{8.1b}$$

Here the product SdE becomes defined more generally, as it includes also the antisymmetric strain and stress parts (see: Chap. 2):

$$\begin{aligned} dU &\to \frac{1}{2}SdE = \frac{1}{2}\sum_{i,k} S_{ik}dE_{ik} \\ &= \frac{1}{2}\left(\frac{1}{3}\sum_{s=1} S_{(ss)}\sum_{k=1} dE_{(kk)} + \sum_{k,l}\hat{S}_{(kl)}d\hat{E}_{(kl)} + \sum_{k,l}\breve{S}_{[kl]}d\breve{E}_{[kl]}\right) \end{aligned} \tag{8.1c}$$

where this expression splits into three products related to the axial, deviatoric shear and rotation strains and stresses (as, e.g., $\delta_{kl}\frac{1}{3}\sum_{s=1}^{3} S_{(ss)}d\hat{E}_{(kk)} = 0$, where $\hat{E}_{(kk)} = 0$); the last product, $\sum_{k,l}\breve{S}_{[kl]}d\breve{E}_{[kl]}$, relates to the rotational components.

The second law of thermodynamics is related to the entropy change:

$$\frac{d\Xi^T}{dt} \geq 0, \quad d\Xi^T = \frac{dQ}{T} \tag{8.2a}$$

A total entropy, $d\Xi^T$, splits into its change, $d\Xi^B$, resulting from a transfer of entropy across the boundaries of a system and an increase of entropy produced within a system, $d\Xi$:

$$d\Xi^T = d\Xi^B + d\Xi \tag{8.2b}$$

However, we can put $d\Xi^B = 0$; then we obtain:

$$\frac{d\Xi^T}{dt} \geq 0 \to \frac{d\Xi}{dt} \geq 0, \quad Td\Xi \geq dQ \tag{8.2c}$$

The basic formulae of thermodynamics remain formally not changed, however the product, SE, becomes changed (cf., 8.1b):

$$SE = \sum_{i,k} S_{ik}E_{ik} = \frac{1}{3}\sum_{s=1} S_{(ss)}\sum_{k=1} E_{(kk)} + \sum_{k,l}\hat{S}_{(kl)}\hat{E}_{(kl)} + \sum_{k,l}\breve{S}_{[kl]}\breve{E}_{[kl]} \tag{8.3}$$

Similarly, for the internal energy and enthalpy we have :

- internal energy:

$$U = Q + \frac{1}{2}(SE)V \tag{8.4a}$$

- enthalpy:

$$H = U - (SE)V \tag{8.4b}$$

Further we recollect some other definitions:

- Helmholtz free energy, F, (having a minimum at the equilibrium state) and the Gibbs free energy, G:

$$F = U - T\Xi\ , \text{ and } G = H - T\Xi \tag{8.4c}$$

where T means the absolute temperature.

We come to the related thermodynamic functions; the plastic deformations depend, among others, on the confining stresses (pressure), however, further on, we will consider the deformations under a constant volume. For the work (dW) done on a body (per volume V) we can write:

$$dW = VS_{(..)}\, dE_{(..)} + VS_{[..]}\, d\bar{E}_{[..]} \geq 0 \tag{8.5}$$

where $S_{(..)}$ and $S_{[..]}$ mean the symmetric and antisymmetric parts of stresses.

For the shear and rotation deformation processes (at a constant pressure), the first law of thermodynamics appears in the form:

$$dU = VdQ + VS_{(..)}\, dE_{(..)} + VS_{[..]}\, d\bar{E}_{[..]} \tag{8.6}$$

where dQ is the heat received in an exchange with the surrounding.

Hence, the internal energy, U, of a body becomes

$$U = Q + \frac{1}{2} VS_{(..)}\hat{E}_{(..)} + \frac{1}{2} VS_{[..]}\bar{E}_{[..]} \tag{8.7}$$

Equilibrium states are characterized by extreme values (minima) of the enthalpy, Helmholtz free energy H, and Gibbs free energy, G:

$$H = U - \frac{1}{2} VS_{(..)}E_{(..)} - \frac{1}{2} VS_{[..]}\bar{E}_{[..]}, \quad F = U - T\Xi\,, \quad G = H - T\tilde{S} \tag{8.8}$$

where T is the absolute temperature and Ξ is the entropy.

From the second law of thermodynamics (8.2a), $d\Xi = dQ/T \geq 0$, we obtain for the irreversible processes an entropy increase.

An important role in any solid continuum is due to defects (Chap. 5). The point defects may be related directly to the electric and magnetic fields and polarization effects.

The Gibbs energy, G, relates to the vacancies, here to the n point defects:

$$G = G_0 + ng^f - TS_c \tag{8.9a}$$

where G_0 is the Gibbs free energy, g^f means the defect formation energy, T the temperature, and S_c the configuration entropy.

At an equilibrium for a constant pressure and temperature we obtain

$$\partial G/\partial n|_{P,T} = 0 \tag{8.9b}$$

and

$$n_{eq} \approx N\exp(-g^f/kT) \tag{8.9c}$$

The $cB\Omega$ model, determining the Gibbs formation energy, has been introduced by Varotsos and Alexoloulus (1986)—see further on:

$$g^f = cB\Omega \tag{8.10a}$$

where c depends only on matrix material and type of the point defects (e.g. Schottky or Frenkel).

Here, we should remind the meaning of the introduced parameters:

$$\Omega = V/N \tag{8.10b}$$

$$\beta = \frac{1}{V}\frac{\partial V}{\partial T}\bigg|_P, \quad B = -V\frac{\partial P}{\partial V}\bigg|_T, \quad \beta B = \frac{\partial P}{\partial T}\bigg|_V, \quad B\Omega = -V\frac{V}{N}\frac{\partial P}{\partial V}\bigg|_T \tag{8.10c}$$

where $\beta B = -\frac{\partial V}{\partial T}\big|_P \frac{\partial P}{\partial V}\big|_T$.

We may note that for any physical quantity, Q, we can write:

$$\frac{\partial Q}{\partial T}\bigg|_V = \frac{\partial Q}{\partial T}\bigg|_P + \beta B\frac{\partial Q}{\partial P}\bigg|_T \tag{8.10d}$$

8.3 Thermodynamics of Point Defects in Solids; The cBΩ Model

A rigorous foundation of the thermodynamic parameters that govern the formation and/or migration of defects in solids has been made by Varotsos and Alexopoulos in a series of publications mainly in Physical Review B during the 1970s and the beginning of 1980s. The main results of these publications have been compiled in a monograph (Varotsos and Alexopoulos 1986).

Following this monograph, the defect formation parameters for example, are defined by comparing a real (i.e., containing defects) crystal either to an *isobaric* ideal (i.e., *not* containing defects) crystal, or to an *isochoric* ideal crystal. Thus, there are two different families of defect formation parameters, which are interconnected through thermodynamic relations.

The *isobaric* parameters are defined in terms of the corresponding Gibbs energy (g^f) as follows (symbols P, T stand for the pressure and temperature, respectively):

$$s^f = -\frac{dg^f}{dT}|_P \tag{8.11a}$$

$$h^f = g^f - T\frac{dg^f}{dT}|_P \text{ and hence } h^f = g^f + Ts^f \tag{8.11b}$$

$$\upsilon^f = \frac{dg^f}{dP}|_T \tag{8.11c}$$

where s^f, h^f and υ^f designate the defect formation entropy, enthalpy and volume, respectively.

When a single mechanism is operating, the self-diffusion activation process is described in terms of the activation Gibbs energy g^{act}, which is the sum of the Gibbs energies g^f and g^m related to the formation and migration processes, respectively. The activation entropy s^{act} and the activation enthalpy h^{act} are then defined accordingly,

$$s^{act} = -\frac{dg^{act}}{dT}|_P \text{ and } h^{act} = g^{act} + Ts^{act} \tag{8.12a}$$

and the diffusion coefficient D is given by

$$D = \gamma\alpha^2 v \exp\left(-\frac{g^{act}}{k_B T}\right) \tag{8.12b}$$

where γ is a numerical constant depending on the diffusion mechanism and the structure, α stands for the lattice constant, and v the attempt frequency which is of the order of the Debye frequency v_D. The symbol k_B stands for the usual Boltzmann constant.

In addition, Varotsos and Alexopoulos, in a series of publications, e.g., Varotsos (1976), Varotsos and Alexopoulos (1977), Varotsos et al. (1978), introduced an interrelation between the defect Gibbs energy g^i (where the superscript i stands for the corresponding process, i.e., defect formation, f, migration, m, or self-diffusion activation, act) and bulk properties in solids. In particular, they proposed a celebrated model - hereafter called cBΩ model—according to which g^i should be proportional to the isothermal bulk modulus B and the mean volume Ω per atom.

The interest on this model that interrelates point defect parameters with macroscopic elastic properties (and this is why it is alternatively termed "elastic" model) has been recently renewed in view of the following findings in diverse fields: First, the model was found by Varotsos et al. (1999) to be consistent with the parameters that describe the time-dependent polarization arising when changing the rate of uniaxial stress in an ionic crystal or by the indenter penetration into its surface. This plays an important role in clarifying the generation mechanism of transient electric signals termed Seismic Electric Signals (Varotsos and Alexopoulos 1984a, b, Varotsos and Lazaridou 1991; Varotsos et al. 1993) that are observed before earthquakes. Second, in high Tc-superconductors, and in particular when doping $YBa_2Cu_3O_{7-\delta}$ with alkaline earth elements, it was found by

Su et al. (2004) that the formation energy for a Schottky defect is compatible with the expectations of the cBΩ-model. Third, the "elastic" model seems to provide a challenging basis for explaining the non-Arrhenius temperature dependence of the viscosity of the glass forming liquids upon approaching the glass transition, e.g., see Dyre (2006), whose full understanding still remains elusive.

Let us now summarize the cBΩ model, in which the defect Gibbs energy g^i is interconnected with the bulk properties of the solid through the relation

$$g^i = c^i B\Omega \tag{8.13}$$

where c^i is dimensionless which—to a first approximation—can be considered as independent of temperature and pressure. The superscript i in Eq. (8.13) refers, as mentioned, to the defect processes under consideration, i.e., f, act, m (formation, self-diffusion activation and migration, respectively), and of course $c^f \neq c^m \neq c^{act}$.

By inserting Eq. (8.13) into Eqs. (8.11a–c), we find

$$s^i = -c^i\Omega(\beta B + \frac{dB}{dT}|_{\mathrm{P}}) \tag{8.14a}$$

$$h^i = c^i\Omega(B - T\beta B - T\frac{dB}{dT}|_{\mathrm{P}}) \tag{8.14b}$$

$$\upsilon^i = -c^i\Omega(\frac{dB}{dP}|_{\mathrm{T}} - 1) \tag{8.14c}$$

where β is the thermal (volume) expansion coefficient. These equations reflect that the ratios s^i/h^i, υ^i/h^i and υ^i/g^i depend *solely* on bulk properties, i.e.,

$$\frac{s^i}{h^i} = -\frac{\beta B + \frac{dB}{dT}|_{\mathrm{P}}}{B - T\beta B - T\frac{dB}{dT}|_{\mathrm{P}}} \tag{8.15a}$$

$$\frac{\upsilon^i}{h^i} = \frac{\frac{dB}{dP}|_{\mathrm{T}} - 1}{B - T\beta B - T\frac{dB}{dT}|_{\mathrm{P}}} \tag{8.15b}$$

$$\frac{\upsilon^i}{g^i} = \frac{1}{B}(\frac{dB}{dP}|_{\mathrm{T}} - 1) \tag{8.15c}$$

Furthermore, the self-diffusion coefficient D can be determined—on the basis of the cBΩ model—at any temperature and pressure from a single measurement: By inserting Eq. (8.13) into Eq. (8.12b) we get

$$D = \gamma\alpha^2 v \exp(-\frac{c^{act}B\Omega}{k_B T}) \tag{8.16}$$

Let us focus on the temperature variation of D at constant pressure. If the value D_1 has been found experimentally at a temperature T_1, the value of c^{act} can be determined because the pre-exponential factor $\gamma\alpha^2 v$ is roughly known. Even if an error of a factor of 2 is introduced by setting ν equal to ν_D, the value of c^{act} remains practically the same. Hence, once the value of c^{act} has been determined from D_1, the value of D_2 for a temperature T_2 can be found through Eq. (8.16) if the elastic data and the expansivity data are available for this temperature.

In what remains in this section, we refer to some recent applications of the cBΩ model in a variety of cases:

First, let us consider an example of the calculation of the heterodiffusion coefficient D for carbon diffusing in α-Fe, see Varotsos (2007a) and references therein, which is a case of major practical interest. The experimental data show that, in the temperature range from 233.9 to 1058 K, the value of D increases by more than 15 orders of magnitude. The adiabatic bulk modulus (B_S) values have been measured in the region 298 1173 K and we convert them to the corresponding B ones - in the standard thermodynamic manner - by using the expansivity and specific heat data given in the literature. The determination of c^{act} is now made at the lowest temperature $T_1 = 234$ K at which D has been measured. For this temperature, the values $D_1 = 5.70 \times 10^{-21}$ or 5.56×10^{-21} cm^2/s have been measured, whereas the expansivity and elastic data mentioned above indicate that $\Omega = 11.75 \times 10^{-24}$cm^3 and $B = 167.2$ GPa. Finally, we assume that the diffusion proceeds by interstitials in octahedral sites and hence we set $\gamma = 1/6$. As for the attempt frequency,ν, we consider that for a given matrix and mechanism, it depends roughly on the mass of the diffusant and, in absence of a better information, we rely on the usual approximation

$$\frac{\nu}{\nu_D} = \left(\frac{m_m}{m_j}\right)^{1/2} \tag{8.17}$$

where m_m and m_j denote the mass of the matrix (m) and the diffusant (j), respectively, and ν_D is $\sim 9.8 \times 10^{12}$s^{-1}. By using these values in Eq. (8.12a), we find

$$c^{act} = 0.06697 \tag{8.18}$$

By inserting the above value of c^{act} into Eq. (8.12a) and using the appropriate data of B and Ω, we compute D for every temperature. For example, at the highest temperature $T = 1058$ K—by using the values $B = 127.5$GPa and $\Omega = 12.18 \times 10^{-24}$cm^3—we compute $D = 2.13 \times 10^{-6}$cm^2/s, which is in satisfactory agreement with the measured experimental value $D = 1.76 \times 10^{-6}$cm^2/s or $D = 2.11 \times 10^{-6}$cm^2/s. Recall that this agreement is achieved although the D value at $T = 1058$ K exceeds that at $T = 234$ K by 15 orders of magnitude.

Second, Varotsos (Varotsos 2007b) investigated the validity of the cBΩ model in diamond, which exhibits certain properties that differ significantly from other materials since it has a very large Debye temperature, $\Theta_D \approx 2246$ K, making it a "quantum" crystal even at room temperature. In general, the point defect parameters of diamond are of great interest in diverse fields. For example, in Earth Sciences its diffusion properties have a specific importance, because natural diamonds and their mineral inclusions provide information about the geochemical character and geotherm of the ancient continental lithosphere. In particular, the spatial distribution of carbon and nitrogen isotopes in diamond provides information on mantle residence time. Diamonds with long residence at high temperature will gradually lose their initial zoning patterns due to diffusion and hence the diffusion data can constrain the maximum possible age of diamonds. Rare diamonds originating from the mantle transition zone should have developed a length scale of $\approx$ 1 mm of isotopic zoning over the age of the Earth (4.5×10^9years). A large body of data on diamond has been accumulated during the last years, which allowed the study of Varotsos (2007b). In particular Varotsos (2007b) made use of the following data: First, the B-values up to 1800 K that were in excellent agreement with experimental results up to 1600 K (based on Brillouin scattering measurements). Second, the data resulting from microscopic calculations which showed that the vacancy formation enthalpy is $h^f = 7.2$ eV and the vacancy formation entropy is $s^f = 2.85k_B$ with a plausible uncertainty which is less than $\pm\, 0.3k_B$.

Varotsos (2007b) proceeded to a numerical check of the cBΩ model as follows: Eqs. (8.9a, 8.9b, 8.9c) for $i = f$ reveal:

$$\frac{s^f}{h^f} = -\frac{\beta B + \frac{dB}{dT}\big|_P}{B - T\beta B - T\frac{dB}{dT}\big|_P} \tag{8.19}$$

The calculation was made for the highest temperature T = 1600 K at which one can rely on the experimental B-values. A least squares fit to a straight line of these B-values as a function of T, gives, in the range $T \geq 1200$ K, the value $\frac{dB}{dT}\big|_P \approx -2.5 \times 10^{-2}$ MPa/K (with a plausible uncertainty of around 10 %), while the B-value (at 1600 K) is $\approx$ 415 GPa. We take also into account that the volume thermal-expansion coefficient for $T = 1600$ K is $\beta \approx 17.26 \times 10^{-6}$ K^{-1}. By inserting these values into Eq. (8.19) we find

$$\frac{s^f}{h^f} \approx 4 \times 10^{-5}\ \mathrm{K}^{-1} \tag{8.20a}$$

with a plausible uncertainty of around 10 %. We now compare this result from the cBΩ model with the value

$$\frac{s^f}{h^f} \approx 3.4^{+0.4}_{-0.4} \times 10^{-5}\ \mathrm{K}^{-1} \tag{8.20b}$$

deduced when inserting the published parameters $s^f = (2.85 \pm 0.3)k_B$ and $h^f = 7.2$ eV mentioned above. This comparison indicates more or less a satisfactory agreement, if one also considers the uncertainties involved. (Note that all the quantities used in the aforementioned calculation come from the articles cited by Varotsos (2007b).)

Third, let us now consider the application of the cBΩ model to PbF_2. The following are the main reasons that explain why the investigation of PbF_2, in general, has attracted a strong interest. Intense experimental efforts have been directed towards finding a candidate material for scintillating detectors to be used in high energy physics experiments in which the particle energy exceeds GeV. An ideal scintillating material, beyond its relatively inexpensive growth process, should be of short radiation length, short decay time, high light yield and a good radiation hardness. Most of these requirements seem to be satisfied by lead fluoride. Furthermore, this material shows a high superionic conductivity with low superionic transition temperature suitable for solid state batteries.

PbF_2 crystallizes from its molten form in the β-phase (cubic, fluorite type structure) and undergoes a pressure-induced phase transition at about 0.4 GPa to the α-phase (orthorhombic cotunnite type structure). In short, the α-phase is a strong scintillator in detector technology, while the β-phase exhibits high superionic conductivity with a low superionic transition temperature. The cBΩ model has been applied succesfully to both phases, i.e., to α-phase and β-phase in Varotsos (2007c) and Varotsos (2008), respectively. In both phases the dominant point defects are anion Frenkel pairs and ionic transport can be successfully described through migration of anion vacancies and interstitials. Among the variety of measurements, the study of the conductivity under pressure is of chief importance since it revealed that the formation volume v^f of the defects is only a small fraction of the molar volume, thus pointing to formation of Frenkel (and *not* Schottky) defects. Varotsos (2007c) showed that this small v^f value as well as the small volumes found for the defect motion, are interconnected with the equation-of-state obtained from a high pressure study of α- PbF_2 by means of angular dispersive synchrotron x-ray powder diffraction techniques

In other words, Varotsos (2007c) showed that the thermal expansivity and the elastic data of α- PbF_2 when employing the cBΩ model leads to values of the Frenkel defects formation volume and the migration volumes for the fluorine vacancy and fluorine interstitial that are comparable to those deduced from conductivity measurements under pressure. Similar agreement between the experimental values and the calculated ones was found by Varotsos (2008) applying the cBΩ model to β- PbF_2.

The cBΩ thermodynamical model as described above is supported by a number of empirical relations very well coinciding with this model; on this basis we arrive at a number of relations important in a search for the earthquake precursors discussed in the next Chap. 9. The thermodynamical model, cBΩ, interconnects the defect parameters with bulk properties. We should keep in mind that periclase (MgO) is one of the major earth-forming minerals prevailing in the lower mantle, where diffusion is considered to be the dominant deformation mechanism and thus is of great importance for geophysics. A number of related papers deal with the estimation of self or hetero diffusion coefficients in MgO under lower-mantle pressure and temperature conditions. Just by means of this thermodynamical model we can join the related defect parameters with its bulk properties. All the related calculations of the diffusion coefficients span a wide range of values and are obtained from macroscopic data (elastic and expansivity data); they compare favourably well with experimental and theoretical coefficients deduced from the microscopic methods.

8.4 Linear Defects

Recall that in relation to the Thermodynamics of the Point Defects we should consider the Schottky defects (cation and anion vacancies at some distance, that do not lie in neighboring sites) in the ionic crystal and the Frenkel defect in a monoatomic crystal (one cation vacancy and that interstitial cation at a remote site). These defects might produce some disorder in structure and additional effects. The earthquake thermodynamics and related electro-magnetic phenomena can be described on these defects and on the dislocation interactions and fracture processes (cf., Chap. 5 and Chap 9).

We may apply such an approach to the linear defects and, instead of active volume, we may consider the active surface related to defect line. Thus, besides of the cBΩ thermodynamical model related to the point defects we may consider the $g^F \rightarrow C\mu_0\lambda_0\Lambda_0^2$ superlattice model for the linear defects (Teisseyre 2001); the related definition (see farther on) may help us to estimate the defect formation energy, g^F, for linear defects.

Here, we introduce a simplification relating to a dislocation distribution as described by the dislocation lines being oriented approximately in the 3 main directions;

$\Lambda_0 = \frac{1}{3}\{\Lambda_1, \Lambda_2, \Lambda_3\} \ll \lambda_0$.

We assume that the considered coordinate system is oriented along the main concentrations of the dislocation lines, or neglect those dislocations whose orientations much deviate from the coordinate directions. The remaining dislocations are those oriented more or less close to the coordinate directions. When we confine ourselves only to a dislocation density related to one given direction, $\alpha^0 = \frac{b}{\Lambda_0^2}$ (b being the Burgers vector), then we may consider a dislocation superlattice with a distance parameter, Λ_0, much greater than a given regular lattice parameter, λ_0; thus $\Lambda_0 \gg \lambda_0$. We should note that to define such superlattice, Λ_0, we had to add the real dislocations, n, some vacant dislocations, $\hat{n}$, to obtain an approximation defining a regular superlattice:

$$N = n + \hat{n} \tag{8.21a}$$

We may relate such dislocation superlattice to the definition of formation Gibbs energy, g^F, given as follows:

$$g^F = \sum \int g^f \frac{\partial l_k}{V} \rightarrow g^F = C\mu_0\lambda_0\Lambda_0^2 \tag{8.21b}$$

This defintion needs, of course, some explanations basing on the relations (8.10a):

- we may have the orientations of the dislocation lines and planes in all possible directions, thus we should introduce a related sum;
- we should join the point defects into a line, thus, we introduce the integrations;

- we integrate the expression, $\partial P/\partial V$, along one of the multitude of dislocation orientations:

$$\int \frac{\partial P}{\partial V}\bigg|_T \partial l_n \approx \int \frac{\partial^3 P}{\partial l_n \partial l_{n+1} \partial l_{n+2}}\bigg|_T \partial l_n \approx \frac{\partial^2 P}{\partial l_{n+1} \partial l_{n+2}}\bigg| \tag{8.21c}$$

- from this relation and from Eqs. (8.21b) and (8.10a, 8.10b, 8.10c) we get

$$g^F = \sum \int g^f \frac{\partial l_k}{V} = c \sum \int B\Omega \frac{\partial l_k}{V} = c \sum \int \frac{\partial P}{\partial V}\bigg|_T \frac{V}{\hat{N}} \partial l_k = c \sum \frac{\partial^2 P}{\partial l_{k+1} \partial l_{k+2}} \frac{V}{\hat{N}} \tag{8.21d}$$

- from Eq. (8.10c) $B\Omega = -V \frac{V}{N} \frac{\partial P}{\partial V}\Big|_T$ and from $\frac{V}{N} \approx \lambda_0 \Lambda_0^2$ we finally obtain:

$$g^F = c \sum \frac{\partial^2 P}{\partial l_{k+1} \partial l_{k+2}} \frac{V}{\hat{N}} \rightarrow g^F = C \mu_0 \lambda_0 \Lambda_0^2 \tag{8.22}$$

where we have introduced a mean value of $\lambda + 2\mu$ as $3\mu_0$; C is a final constant.

In this way we may arrive at the formation Gibbs energy related to a superlattice, $g^f \rightarrow g^F$, as already presented (cf., Eq. 8.11b).

Further, we can define the dislocation lattice density, $\alpha^0 = \frac{b}{\Lambda_0^2}$, related to the real density of dislocations, α:

$$\alpha^0 = \frac{b}{\Lambda_0^2}, \quad \alpha = \alpha^0 - \hat{\alpha} = \frac{b}{\Lambda_0^2}\left(1 - \frac{\hat{n}}{\hat{N}}\right) \tag{8.23}$$

where Λ a means a unit value of the superlattice, $\hat{\alpha}$ and $\hat{n}$ being the density and number of the vacant dislocations.

For the related formation Gibbs energy, g^F, and Gibbs energy, G, we write:

$$G = G^0 + ng^F - T\Xi^F \tag{8.24}$$

where G^0 is the Gibbs free energy of perfect isobaric crystal, g^F is the formation Gibbs energy and Ξ^F is the related configuration entropy.

Thus, besides the point defects we consider also the linear defects which may be more directly related to the stress load and its concentration; however, such processes forming the linear defects induce also an important appearence of point defects, more directly related to the electric, magnetic, or electric polarization precursory signals; see the next chapter. A direct connection between these linear and point defects should be underlined; such an approach might form a good background for a search of efficient earthquake precursors.

Both defects, point and linear, are mutually combined; we may assume that the linear defects increase in an earthquake preparation zone and this process activates the point defects and related emitted signals. The point and linear defects are subjected to similar thermodynamical rules.

We should also recall the local dislocation balace law (cf., Teodosiu 1970; Teisseyre 1990a, b, 1995, and 2001):

$$\frac{\partial \alpha_{kl}}{\partial t} + \sum_{m} \frac{\partial(\alpha_{kl} V_m - \alpha_{ml} V_k)}{\partial x_m} = 0 \tag{8.25}$$

where V_k means the dislocation velocity vanishing along the dislocation line; for the relation between the dislocation velocity and stresses, see: Mataga et al. (1987) and Teisseyre (2001).

Further, worth mentioning here are also the dislocation/stress relations (Chap. 5; cf., Teisseyre 2001, 2006, and 2008a, b).

We remind that the isobaric defect formation relates to the pressure and temperature, P and T, while isochoric defect formation to the volume and temperature, V and T, constants.

8.5 Superlattice and Shear Band Model

It is to be kept in mind that the formation of a defect line, e.g., dislocation, leads to the negative contribution to the Gibbs free energy, hence, we cannot reach minimum for any number of the dislocations (Kocks et al. 1975). To omit this problem, we introduce the concept of the dislocation superlattice. Important considerations on the equilibrium thermodynamics, irreversible processes, and local equilibrium (Prigogine 1976) as applied to the dislocation superlattice have been given by Teisseyre and Majewski (2001a, b). The ideal superlattice, a cubic lattice with the equal distances, Λ, between the dislocations, $\Lambda_0 \gg \lambda_0$ (here, λ_0 relates to the regular particle lattice), could be treated as a reference equilibrium state. We shall note, however, that a shear band model (Teisseyre and Majewski, 1990, 1995a, b, 2001a, b; Majewski and Teisseyre, 1997, 1998, 2001; see Fig. 8.1) may be treated only as a very rough approximation of a real situation. To approach such an ideal superlattice, we start with a real dense network of dislocations and then define a superlattice which in reality forms a more or less chaotic dynamic structure. To define an approximation of a superlattice we add to a real dislocation structure a suitable number of the vacant dislocations in order to obtain a distribution of dislocations and vacant dislocations with a minimal departure from the ideal superlattice. This may be achieved only when considering a state with a dense network of dislocations: due to the interactive forces between the dislocations their any real distribution is not quite chaotic; therefore, the proposed procedure seems to be realistic. A minimum departure from the ideal distributions defines the number and distribution of the vacant dislocations (Kuklińska 1996).

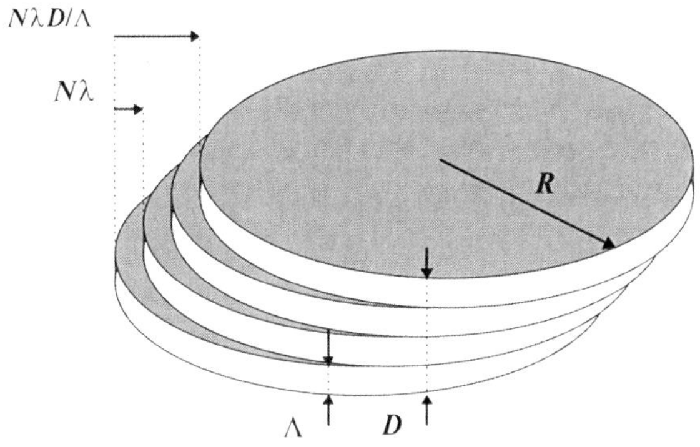

Fig. 8.1 A shear band model of the earthquake focal region as related to the lattice constant, λ_0 and the superlattice constant, Λ_0

The Gibbs energy of the superlattice can be defined by such a number of the vacant dislocations, which brings us nearer to the equilibrium state (Teisseyre 1969, 1970; Teisseyre and Majewski 1995a, b, 2001a, b). The possible super-Gibbs energy is

$$G = G^0 + Ng^F \tag{8.26a}$$

The line defects considered, N, may be related to different types of dislocations; however, the procedure outlined above refers, rather, to one chosen type of dislocations: the edge dislocations, screw ones, or the disclinations, otherwise called rotational dislocations (Teisseyre and Górski 2009). In the microscopic model we assume that all linear defects bring the discontinuity distribution to the crystal lattice constant, λ_0.

In a material under the influence of a complex load history, the different-type defects appear; the edge and screw dislocations are crossing each other and forming the jogs and kinks on their lines. A kink means a step on a screw dislocation line when a dislocation line and its plane lie along its glide plane, and the jog, being a step out of a glide plane when an edge dislocation line and its plane cross a glide plane perpendicularly. Of course, any defect motion is controlled by the stress resistance (cf., Kocks et al. 1975).

Now, we can consider an equilibrium distribution of the vacant dislocations and disclinations. Only when associating the thermodynamic functions with the vacant dislocations in a superlattice, we can reach and use a minimum of the Gibbs free energy. Our approach follows the Thermodynamics of Point Defects (Varotsos and Alexopoulos 1986), in which a number of important results have been achieved.

As we already pointed out, an approximation of a superlattice with a characteristic distance Λ (superlattice constant) can be achieved when adding to the real defects a suitable number of vacant defects placed in suitable sites. Further, we will consider a cubic superlattice with dislocations orientated in accordance with the shears (such an orientation may be estimated from the earthquake source tensor solution). In a superlattice defined in this way a mean value of distances between defects, Λ_0, follows from the average spacing between the real and vacant defects.

To simplify the theory and model we assume that the related Burgers vectors and the Frank vector are equal to the basic lattice constant (the microscopic approach).

Thus, using the dislocation densities for the edge and screw dislocations, α^E and α^S, and for disclinations, θ^0, (cf., Fig. 8.2) with the respective Burgers vectors, b^E and b^S, and the angular moment related Frank vector, $\lambda_0\beta^R$, (where λ_0 is a regular lattice constant and β^R the rotation angle):

$$\alpha^E = \frac{b^E}{\Lambda_0^2}\ \alpha^S = \frac{b^S}{\Lambda_0^2}\ \theta^0 = \frac{\lambda_0\beta^R}{\Lambda_0^2} \tag{8.27a}$$

Further, we assume that the existing shears produce the equal effects on all types of linear defects, for the edge and screw dislocations and disclinations, $\alpha^E, \alpha^S, \theta^0$; thus, for simplicity we put:

$$\alpha^0 = \frac{\lambda_0}{\Lambda_0^2} \tag{8.27b}$$

with $\alpha^0 = \alpha^E + \alpha^S + \theta^0$, and $b^E = b^S = \lambda_0\,\beta^R \approx \lambda_0$.

This is a standard value; however, due to a shear load and formation of arrays of dislocations, it will adequately grow to some equilibrium value. Moreover, we shall note that a real distribution of defects is not quite chaotic; due to the interactions between these objects there exists some kind of structure. Due to this fact, we can assume that to a real continuum with defects it is possible to add a number of vacant defects in order to approach, with the real and vacant defects, a state presenting a minimal departure from the ideal superlattice. Thus, we add the appropriate number of vacant defects, $\tilde{n}$, to real ones, n, in order to get closer to the conditions of an ideal superlattice. A condition for the network to be best fitting to the ideal superlattice (Kuklińska 1996) is:

$$n + \tilde{n} = N \tag{8.28a}$$

Then, for a real equilibrium density of defects and vacant ones we may write:

$$\alpha^{Eq} = \alpha + \tilde{\alpha} + \theta^0 \approx N\frac{\lambda_0}{\Lambda_0^2}\exp\left(-\frac{g^F}{kT}\right) \tag{8.28b}$$

Note that the dislocation density is a tensor field describing a density of dislocations crossing a given plane; in the above equation we have written the number of dislocations crossing a plane of any orientation. The edge and screw dislocation densities related to a shear field, $S_{(kn)}$, belong the defects appearing on the two planes perpendicular to the versors v_k and v_n, while disclination density (rotational dislocation) of a given sign belongs only to one plane perpendicular to versor v_i, as related to $\frac{1}{2}\varepsilon_{ikn}S_{[kn]}$. In this situation the numbers of the particular defects in the superlattice will theoretically be in the proportions:

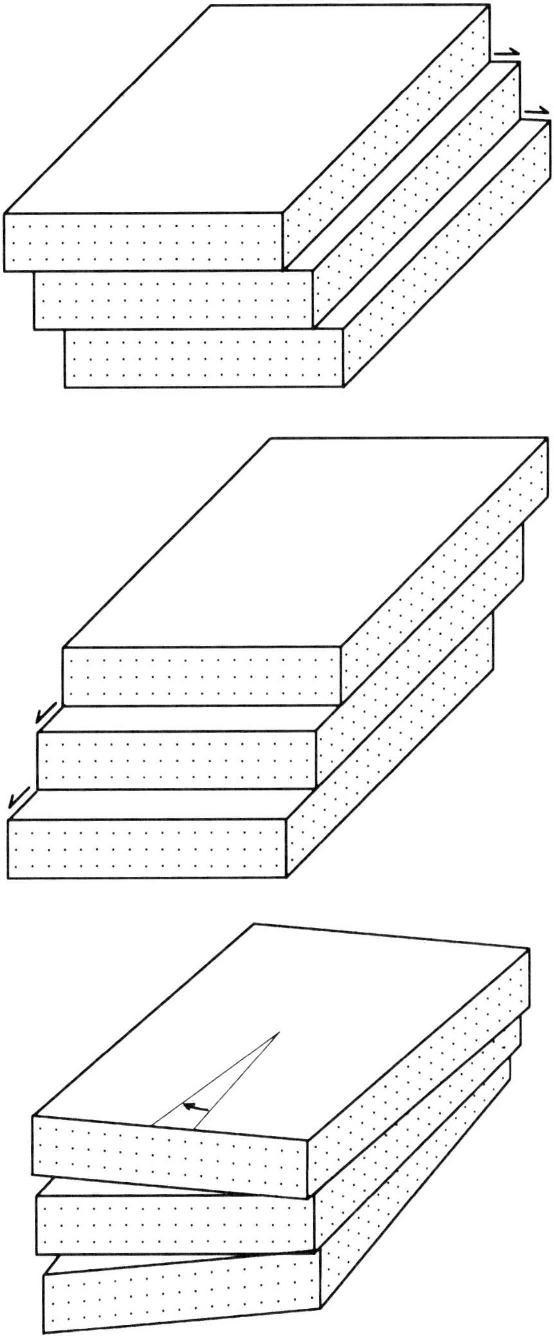

Fig. 8.2 Different linear defects on the sides of the parallelepiped related to the superlattice; from *top*: edge and screw dislocations and disclinations

$$N^E = \frac{2}{5}N, \quad N^S = \frac{2}{5}N, \quad N^R = \frac{1}{5}N \quad \text{at} \quad N = N^E + N^E + N^R \tag{8.29}$$

The same proportions can be assumed for real and vacant defects:

$$n^E = \frac{2}{5}n, \quad n^S = \frac{2}{5}n, \quad n^R = \frac{1}{5}n; \quad \tilde{n}^E = \frac{2}{5}\tilde{n}, \quad \tilde{n}^S = \frac{2}{5}\tilde{n}, \quad \tilde{n}^R = \frac{1}{5}\tilde{n} \tag{8.30}$$

where $n = n^E + n^E + n^R$, and $\tilde{n} = \tilde{n}^E + \tilde{n}^S + \tilde{n}^R$.

And for a density of linear defects and vacant ones, we write

$$\alpha^0 = \alpha^{E0} + \alpha^{S0} + \theta^0; \quad \alpha^{E0} \approx \frac{2}{5}\frac{\lambda_0 N}{\Lambda_0^2}, \quad \alpha^{S0} \approx \frac{2}{5}\frac{\lambda_0 N}{\Lambda_0^2}, \quad \theta^0 \approx \frac{1}{5}\frac{\lambda_0 N}{\Lambda_0^2} \tag{8.31}$$

With the help of the thermodynamic principles, the vacant defect density can be independently given by the corresponding value related to an equilibrium state; it means, when the Gibbs energy $\tilde{G} = \tilde{G}^0 + \tilde{n}\, g^F - T\Xi_c$ (here g^F and Ξ_c mean the Gibbs formation energy for vacant dislocations and configuration entropy, respectively), related to the vacant dislocations, reaches minimum (cf. Teisseyre and Majewski 2001a, b), we come to the equilibrium state (cf., Eqs. 8.15a, 8.15b, 8.15c):

$$\tilde{n}^{Eq} = N\exp\left(-\frac{\tilde{g}^F}{kT}\right) \; ; \text{ or } \tilde{\alpha}^{Eq} = \frac{\lambda}{\Lambda^2}\exp\left(-\frac{\tilde{g}^F}{kT}\right) = \alpha^0\exp\left(-\frac{\tilde{g}^F}{kT}\right), \; \alpha^0 = \frac{\lambda_0}{\Lambda_0^2} \tag{8.32}$$

where g^F is the Gibbs formation energy per length of the crystal lattice, λ_0, which may be different for each kind of defects, but here assumed to be the same for them all; k is the Boltzmann constant, and we assumed a reference density as α^0; Λ_0 is a superlattice constant (a length of displacement in a shear-band model).

In our former model (Teisseyre 1996, 1997, 2001; Majewski and Teisseyre 2001; Teisseyre et al. 2006a,b, 2008), we have assumed that just before an earthquake the number of dislocations would reach a very high value. Close to fracture process the defect arrays are formed; these arrays would destroy the structure superlattice structure. According to the band structure model, the total Burgers vector, b, before a fracture event would increase up to the value (cf., Fig. 8.1):

$$b = n_A \lambda_0 \frac{D}{\Lambda_0} \tag{8.33a}$$

where the number n_A relates to stress concentration at a tip of array; D is the total source thickness, and thus D/Λ_0 is the number of fracture planes in a source.

Therefore, we have an approximate formula for the maximum of defect density before an event:

$$\alpha^0 = \frac{\lambda_0}{\Lambda_0^2} \rightarrow \alpha^M = n_A \lambda_0 \frac{D}{\Lambda_0^3} \tag{8.33b}$$

and therefore, before a fracture the equilibrium density of defects could increase from the number, α^{eq} (ideal superlattice) up to that maximum.

We may note that the obstacles may stop and bend dislocations at the following probability, $\tilde{p}$:

$$\tilde{p} = \exp\left(-\frac{dG}{kT}\right) \tag{8.34}$$

In a fracture process when these defects and the opposite defects and their arrays mutually undergo the annihilation in micro and macro-processes, as during an earthquake event, such a density will drop to a much lower value; in a very rough approximation, to the equilibrium value. In the mentioned model (Majewski and Teisseyre 2001) we have already assumed that starting before an event with dense distribution of defects we may come, after the event, close to the equilibrium value as given by equilibrium number of the vacant defects, defined in relation (8.20a, b); thus, theoretically a fracture process would mean transition from maximal value of defect content to the equilibrium state:

$$\alpha^M \rightarrow \tilde{\alpha}^{eq} = \frac{\lambda}{\Lambda^2} \exp\left(-\frac{\tilde{g}^F}{kT}\right), \tag{8.35}$$

where $\Delta\alpha = n_A \lambda \frac{D}{\Lambda^3} - \frac{\lambda}{\Lambda^2} \exp\left(-\frac{\tilde{g}^F}{kT}\right) = \frac{\lambda}{\Lambda^2}\left(n_A \frac{D}{\Lambda} - \exp\left(-\frac{\tilde{g}^F}{kT}\right)\right)$

In analogy to the Thermodynamics of the Point Defects (Varotsos and Alexopoulos, 1986) we found out that the Gibbs and entropy functions for a body with the line defects become (Teisseyre and Majewski 2001a, b):

$$G = G^0 + \tilde{n} g^F - T\Xi^C, \quad \Xi = \Xi^0 + \tilde{n}\left(\Xi^F + k + \frac{g^F}{T}\right) \tag{8.36}$$

where we refer these function per line vacancy to a number of the vacant dislocations, $\tilde{n}$ (cf., Eq. 8.18), similarly referring to the formation energy for the vacant dislocation; $\Xi^F = -\partial \tilde{g}^F / \partial T$.

The value of the Gibbs formation energy can be adopted using the $C\mu_0\lambda_0\Lambda_0^2$ model (Teisseyre and Majewski 2001a,b and Teisseyre 1996, 1997, 2001). That model has been adopted from the model CBΩ for the point defect thermodynamics (Varotsos and Alexopoulos 1986; cf., Varotsos and Lazaridou 2001), where:

$B = -V \frac{\partial P}{\partial V}|_T$ (isothermal bulk modulus), $\beta = \frac{1}{V}\frac{\partial V}{\partial T}|_P$ (thermal volume expansion),

$\Omega = \frac{V}{N}$ (mean volume per atom)

The $C\mu_0\lambda_0\Lambda_0^2$ model (Teisseyre and Majewski 2001a, b) means that we can define the Gibbs formation energy of a vacant dislocation as related to a lattice constant and a square of the superlattice defect line (cf., Fig. 8.1):

$$g^F = C\mu_0\lambda_0\Lambda_0^2 \tag{8.37}$$

Our further considerations on the earthquake thermodynamics will be based on both models: we may suppose that a concentration of dislocations activate also the point defects being more directly related to the electric and magnetic precursors. In this way, we approach a state just before a macroscopic fracture. Further, we assume that just before an earthquake process the ideal superlattice state is completely destroyed by the linear defect arrays. That means that instead of one defect per supperlattice, Λ_0, there appears an array of defects.

After Roux and Guyon (1985), the Hamiltonian related to their beam model includes the four terms: axial, shear, bending, and torque. For stored rotational energy we may write in (cf., Eq. 8.2a):

$$E^{\mathrm{Rot}} = 2\mu\breve{E}_{sk}\breve{E}_{sk} = \mu\left(\frac{\partial u_k}{\partial x_s}\frac{\partial u_k}{\partial x_s} - \frac{\partial u_s}{\partial x_k}\frac{\partial u_k}{\partial x_s}\right) \tag{8.38}$$

This expression may be interpreted as the total torque minus bending energy; comparing this result with that by Roux and Guyon (1985), we find an essential difference, because here the torque energy as defined by the first term will be diminished by bending relaxation process.

The $C\mu_0\lambda_0\Lambda_0^2$ model, for the linear defects thermodynamics, leads to the following expressions for the equilibrium number and density of the vacant linear defects, $\hat{n}^E, \hat{\alpha}^E$, being related to the Gibbs energy minimum:

$$\tilde{n}^E = N\exp\left(-\frac{C\mu_0\lambda_0\Lambda_0^2}{kT}\right),\ \tilde{\alpha}^E = \frac{\lambda_0}{\Lambda_0^2}\exp\left(-\frac{C\mu_0\lambda_0\Lambda_0^2}{kT}\right) \tag{8.39}$$

We note that, here, the constant C will depend only on the matrix material and the type of defects considered, that may differ for the edge and screw dislocations and disclinations; in our approach we assume that this constant remains the same for the edge and screw dislocations and for disclinations. We repeat that, in a rough approximation, we assume that before an earthquake a crack superlattice may approach the state being almost completely filled-in by the linear defects: in a unit of superlattice, Λ_0, one array of defects appears, that means, in the cube Λ_0^3, we will have three perpendicular arrays just before an earthquake event. This is a new additional assumption to our shear-band model in fracture continuum: one array per superlattice constant, meaning 3 perpendicular arrays in the superlattice cube. After an earthquake, we assume that defect distribution will drop to the state near equilibrium (8.27a, b); this becomes possible due to the annihilation processes in which the positive and negative defects mutual annihilate with an energy release. Such defect coalescence processes lead to a rebound value of seismic moment and energy released. We can exactly follow the results presented by

Teisseyre and Majewski (2001a, b); we arrive at the expressions for seismic moment, energy release, and entropy drop:

$$M_0 = \pi\mu\left(\frac{\Lambda_0}{\lambda_0}\right)^2 R^2 D \exp\left(-\frac{C\mu_0\lambda_0\Lambda_0^2}{kT}\right) \tag{8.40}$$

$$\Delta E = G - G_0 = \frac{\pi R^2 D}{\lambda_0^3} kT \exp\left(-\frac{C\mu_0\lambda_0\Lambda_0^2}{kT}\right) \tag{8.41}$$

$$\Delta\Xi = \frac{\pi R^2 D}{\lambda_0^3} k\left(1 + \frac{C\mu_0\lambda_0\Lambda_0^2}{kT}\right)\exp\left(-\frac{C\mu_0\lambda_0\Lambda_0^2}{kT}\right) \tag{8.42}$$

where the focal region is defined as $\pi R^2 D$ and the energy release is related to a change of the Gibbs free energy and number of coalescences between defects.

However, contrary to that former approach we shall take into account that the released energies contain the symmetric and antisymmetric stresses; thus, when comparing these theoretical expressions with the observation data we include there the energy drops as related to symmetric deviatoric strain, axial strain and rotation field. Therefore, instead of Eq. 8.37, the new model is split into three ones related to the axial, deviatoric and rotation strains:

$$g^F = C\mu_0\lambda_0\Lambda_0^2 \to \left\{\bar{g}^F, \hat{g}^F, \breve{g}^F\right\} = \left\{\bar{C}\mu_0\lambda_0\Lambda_0^2, \hat{C}\mu_0\lambda_0\Lambda_0^2, \breve{C}\mu_0\lambda_0\Lambda_0^2\right\} \tag{8.43}$$

Accordingly, relations (8.28a, 8.28b), (8.29) and (8.30) will split into the three different sets for M_0, ΔE and $\Delta\Xi$:

$$\left\{\bar{M}_0, \hat{M}_0, \breve{M}_0\right\} = \pi\mu\left(\frac{\Lambda_0}{\lambda_0}\right)^2 R^2 D \exp\left(-\frac{\left\{\bar{C}, \hat{C}, \breve{C}\right\}\mu\lambda_0\Lambda_0^2}{kT}\right) \tag{8.44}$$

$$\left\{\Delta\bar{E}, \Delta\hat{E}, \Delta\breve{E}\right\} = G - G_0 = \frac{\pi R^2 D}{\lambda^3} kT\exp\left(-\frac{\left\{\bar{C}, \hat{C}, \breve{C}\right\}\mu\lambda_0\Lambda_0^2}{kT}\right) \tag{8.45}$$

$$\left\{\Delta\bar{\Xi}, \Delta\hat{\Xi}, \Delta\breve{\Xi}\right\} = \frac{\pi R^2 D}{\lambda_0^3} k\left(1 + \frac{\left\{\bar{C}, \hat{C}, \breve{C}\right\}\mu\lambda_0\Lambda_0^2}{kT}\right)\exp\left(-\frac{\left\{\bar{C}, \hat{C}, \breve{C}\right\}\mu\lambda_0\Lambda_0^2}{kT}\right) \tag{8.46}$$

Further, we may note that the relation for seismic efficiency, η, may also differ for the different release processes:

$$\frac{M_0}{E^{radiated}} = \frac{\mu_0\lambda_0\Lambda_0^2}{\eta kT}, \quad E^{radiated} = \eta\Delta E \tag{8.47}$$

where it is justified to include the different efficiencies: $\eta \to \{\bar{\eta}, \hat{\eta}, \breve{\eta}\}$.

We should add that, due Transition State Theory, the thermally activated glide relates to some transition state in which the molecules collide with enough energy to form a transition structure; however, due to the Quantum Theory, there still might be a possibility for an appearance of a tunnel-like micro-glide across a barrier.

Finally, we can add that the fracture transport with the bond breaking and granulation processes proceeds with the bond breaking preceding by $\pi/2$ in phase the rebound rotation motion, and then a massive defect motion occurs and an energy release takes place with defect annihilation. However, to deal with such relations (cf., Eqs. 8.44, 8.45, 8.46), we will need to have the data from the future-expected network of the stations equipped with the array seismometers and rotation sensors, to achieve on this basis the new statistical relations permitting to estimate the observed magnitudes, energy drops and entropy drops, as related to the three models discussed $\{\bar{C}, \hat{C}, \breve{C}\}$. Unfortunately, this is a very far future and just now we may only note that our last considerations are not related to the presently existing seismological network, and therefore we cannot use them in pratice.

In the above considerations, we took into account the processes related to slip and rotation; a total released energy includes that related to wave radiation and that to heat caused by friction. A role of micro-fracturing in the bond breaking process is similar under both the confining pressure and the external shears; however, we observe the essential differences for rotations in larger scales. A confining condition leads to the formation of induced opposite-sense arrays of dislocations, resulting in fragmentation processes and chaotically oriented macro-rotations, leading therefore to a rotation release process, while a shear condition leads to more concentrated fracturing along some planes, high shear strain release and correlated rotations.

References

Dyre JC (2006) Colloquium: The glass transition and elastic models of glass-forming liquids Rev. Mod Phys 78:953–972

Kocks UF, Argon AS, Ashby MF (1975) Thermodynamics and kinetics of slip. Pergamon Press, New York, p 288

Kuklińska M (1996) Thermodynamics of line defects: construction of a dislocation superlattice. Acta Geophys Polon 44:237–249

Majewski E, Teisseyre R (1997) Earthquake thermodynamics. Tectonophysics 277(1–3):219–233

Majewski E, Teisseyre R (1998) Anticrack-associated faulting in deep subduction zones. Phys Chem Earth 23:1115–1122

Majewski E, Teisseyre R (2001) Anticrack-associated faulting and superplastic flow in deep subduction zones. In: Teisseyre R, Majewski E (eds) Earthquake thermodynamics and phase transformations in the Earth's interior. (International Geophysics Series vol 76). Academic Press, San Diego, pp 379–39

Mataga P, Freund L, Hutchinson J (1987) Crack tip plasticity in dynamic fracture. J Phys Chem Solids 48:985–1005. doi:10.1016/0022-3697(87)90115-6
Roux S, Guyon E (1985) Mechanical percolation: a small beam lattice study. J Phys Lett (Paris) 46(21):999–1004
Su H, Welch DO, Winnie W-N (2004) Strain effects on point defects and chain-oxygen order-disorder transition in 123 cuprate compounds. Phys Rev B 70:054517
Teisseyre R (1969) Dislocational representation of thermal stresses. Acta Geophys Polon 16:3–12
Teisseyre R (1970) Crack formation and energy release caused by the concentration of dislocations along fault planes. Tectonophysics 9(6):547–557
Teisseyre R (1990a) Earthquake rebound: energy release and defect density drop. Acta Geophys. Pol XXXVIII(1):15–20
Teisseyre R (1990b) Earthquake Premonitory and Rebound Theory: Synthesis and revision of Principles. Acta Geophys. Pol XXXVIII(3):269–278
Teisseyre R (1995) Deformation and geometry. In: Teisseyre R (ed) Theory of Earthquake Premonitory and Fracture Processes. PWN (Polish Scientific Publishers), Warszawa, pp 504–511
Teisseyre R (1996) Shear band thermodynamical earthquake model. Acta Geophys Pol 44(3):219–236
Teisseyre R (1997) Shear band thermodynamical model of fracturing with a compressional component. In: Gibowicz S, Lasocki S (eds) Rockburst and seismicity in mines. A.A.Balkema, Rotterdam, Brookfield, pp 17–21
Teisseyre R (2001) Shear band thermodynamic model of fracturing. In: Teisseyre R, Majewski E (eds) Earthquake thermodynamics and phase transformations in the Earth's interior. Academic Press, San Diego, pp 279–292
Teisseyre R (2008a) Introduction to asymmetric continuum: dislocations in solids and extreme phenomena in fluids. Acta Geophys Pol 56(2):259–269
Teisseyre R (2008b) Asymmetric continuum: standard theory. In: Teisseyre R, Nagahama H, Majewski E (eds) Physics of asymmetric continuum: extreme and fracture processes: earthquake rotation and soliton waves. Springer, Heidelberg, pp 95–109
Teisseyre R, Górski M (2009) Fundamental deformations in asymmetric continuum. Bull Seism Soc Am 99(2B):1132–1136. doi:10.1785/0120080091
Teisseyre R, Majewski E (1990) Thermodynamics of line defects and earthquake processes. Acta Geophys Pol 38:355–373
Teisseyre R, Majewski E (1995a) Thermodynamics of line defects. In: Teisseyre R (ed) Theory of Earthquake Premonitory and Fracture Processes. PWN, Warsaw, pp 324–332
Teisseyre R, Majewski E (1995b) Earthquake thermodynamics. In: Teisseyre R (ed) Theory of Earthquake Premonitory and Fracture Processes. PWN, Warsaw, pp 586–590
Teisseyre R, Majewski E (2001a) Thermodynamics of line defects and earthquake thermodynamics. In: Teisseyre R, Majewski E (eds) Earthquake thermodynamics and phase transformations in the Earth's interior. Academic Press, San Diego, pp 261–278
Teisseyre R, Majewski E (2001b) Physics of earthquakes. In: Lee WHK, Kanamori H, Jennings PC (eds) International Handbook of Earthquake and Engineering Seismology. Academic Press, San Diego
Teisseyre R, Białecki M, Górski M (2006a) Degenerated asymmetric continuum theory. In: Teisseyre R, Takeo M, Majewski E (eds) Physics of asymmetric continuum: extreme and fracture processes: earthquake rotation and soliton waves. Springer-Verlag, Berlin, Heidelberg, pp 43–55
Teisseyre R, Górski M, Teisseyre KP (2006b) Fracture-band geometry and rotation energy release. In: R Teisseyre, M Takeo, E Majewski (eds) Physics of asymmetric continuum: extreme and fracture processes: earthquake rotation and soliton waves, Springer-Verlag, Berlin, Heidelberg, pp 169–183
Teisseyre R, Górski M, Teisseyre KP (2008) Fracture processes: spin and twist-shear coincidence. In: Teisseyre R, Nagahama H, Majewski E (eds) Physics of asymmetric

continuum: extreme and fracture processes: earthquake rotation and soliton waves. Springer-Verlag, Berlin, Heidelberg, pp 111–122
Teodosiu C (1970) A dynamic theory of dislocations and its applications to the theory of the elastic plastic continuum. In: Simmonds A (ed) Fundamental Aspects of Dislocation Theory, vol. 317(2), National Bureau of Standards Special Publication, pp 837–876
Varotsos P (1976) Comments on the formation entropy of a Frenkel defect in BaF2 and CaF2. Phys. Rev. B 13:938
Varotsos P (2007a) Self-diffusion in sodium under pressure revisited. J Phys: Condens Matter 19:176231
Varotsos P (2007b) Calculation of point defect parameters in diamond. Phys. Rev. B 75:172107
Varotsos P (2007c) Defect volumes and the equation of state in α-PbF2. Phys Rev B 76(4):092106
Varotsos P (2008) Point defect parameters in b-PbF2 revisited. Solid State Ionics 179:438–441
Varotsos P, Alexopoulos K (1977) Calculation of the formation entropy of vacancies due to anharmonic effects. Phys. Rev. B 15:4111–4114
Varotsos P, Alexopoulos K (1984a) Physical properties of the variations of the electric field of the earth preceding earthquakes, I. Tectonophysics 110:73–98
Varotsos P, Alexopoulos K (1984b) Physical properties of the variations of the electric field of the earth preceding earthquakes, II. Determination of epicenter and magnitude, Tectonophysics 110:99–125
Varotsos P, Alexopoulos K (1986) Thermodynamics of Point Defects and their relation with the bulk properties, North Holland, Amsterdam.
Varotsos P, Lazaridou M (1991) Latest aspects of earthquake Prediction in Greece based on Seismic Electric Signals. I, Tectonophysics 188:321–347
Varotsos P, Ludwig W, Alexopoulos K (1978) Calculation of the formation volume of vacancies in solids. Phys. Rev. B 18:2683–2691
Varotsos P, Alexopoulos K, Lazaridou M (1993) Latest aspects of earthquake prediction in Greece based on Seismic Electric Signals II. Tectonophysics 224:1–37
Varotsos P, Sarlis N, Lazaridou M (1999) Interconnection of defect parameters and stress-induced electric signals in ionic crystals. Phys. Rev. B 59:24–27
Varotsos PA, Lazaridou M (2001) Thermodynamics of point defects. In: Teisseyre R, Majewski E (eds) Earthquake thermodynamics and phase transformations in the Earth's interior. Academic Press, San Diego, pp 231–259

Chapter 9
Mutual Interactions: Electric/Magnetic Fields and Strains

9.1 Introduction

In this chapter we discuss interactions of the mechanical strain fields with the electric and magnetic fields; our intention was to achieve general forms of such interactions. This task may play a very important role in different problems, e.g., in the search of earthquake precursors. We present here an attempt to include a counterpart of the defects and a role of thermodynamical description of interaction processes. We include here the basic studies and achievements of the Professor P. Varotsos and his group exposing an essential role both for the thermodynamical aspects and in the transient variations of the electric field of the Earth as observed before the occurrence of major earthquakes. The achievements presented here give an important basis leading to a general form of the discussed interaction schema; that is, we present some general relations joining the electric and magnetic fields and also the electric polarization with the strains and deformations (e.g., the axial, shear and rotational strains).

The mutual interactions between strains and electric/magnetic fields, including the electric polarizations, play an important role the in the thermodynamical problems and in many physical phenomena and processes, e.g., in a search of the earthquake precursors. We include here a counterpart of the defects and a role of thermodynamical description in the related interaction processes. We intend to include here the thermodynamical aspects of these interactions and a related role of the defects. We will consider also some physical aspects of these interaction processes, e.g., related to the precursory earthquake phenomena.

Interaction between strains and electric/magnetic fields is very important in a search for the earthquake precursors. However, we may notice at least two different assumptions; one relates to the thermodynamics of the point defects and possible appearence of the electric or magnetic precursors, while another approach is based on the linear defects and their role in a thermodynamical description of the interaction processes.

R. Teisseyre and M. Teisseyre-Jeleńska, *Asymmetric Continuum*,
GeoPlanet: Earth and Planetary Sciences, DOI: 10.1007/978-3-642-31860-3_9,

The point defects and related interactions with strains before an earthquake event are considered in a number of important papers: here, first of all, we should take into account the numerous papers and studies by Professor P. Varotsos's group (cf., Varotsos et al. 2001, 2006a, b). Basing on the thermodynamical properties of point defects (cf., Chap. 8) and on a number of studies of the precursory events, let us present the main achievements related to this important approach.

We first describe some main achievements related to the interaction problems.

9.2 Earthquake Prediction: Seismic Electric Signals and Natural Time Analysis

Varotsos and Alexopoulos have shown that when the pressure (stress) reaches a *critical* value, a *cooperative* orientation of the electric dipoles existing in the solid may occur, which results in the emission of a transient electric signal. This may happen before an earthquake since the stress gradually increases in the focal region before the rupture. Along this direction, a detailed experimentation started in Greece in 1981, which showed (Varotsos and Alexopoulos 1984a, b) that transient variations of the electric field of the Earth are actually observed before the occurrence of major earthquakes. These signals are termed *Seismic Electric Signals*, SES, of the so-called VAN earthquake prediction method; VAN comes from the initials of Varotsos, Alexopoulos and Nomikos.

The main properties of SES could be summarized as follows (Varotsos and Alexopoulos 1984a, b): First, the SES amplitude is interrelated with the magnitude of the impending earthquake. This interrelation is in fact a power-law which (since 1984) corroborates the notion that the approach to a critical point (second order phase transition) is accompanied by fractal structure, in accordance with the aforementioned SES generation mechanism proposed by Varotsos and Alexopoulos, since *criticality* is always accompanied by *fractality*. Second, SES cannot be observed at all points of the Earth's surface but only at certain points called "sensitive points". Each sensitive station enables the collection of SES only from a restricted number of seismic areas (*selectivity effect*). A map showing the seismic areas that emit SES detectable at a given station is called "selectivity map of this station". This allows the determination of the epicenter of an impending earthquake and the SES electric field precedes markedly (~ 1s) the time-derivative of the relevant magnetic field (Varotsos and Lazaridou 1991; Varotsos et al. 1993). Third, at epicentral distances of the order of 100 km, field variations (Varotsos et al. 2003).

The physical properties of SES can be theoretically explained, if we take into account the aforementioned SES generation mechanism together with the existence of inhomogeneities in the Solid Earth's Crust (for a review see Varotsos 2005). The SES collection from the real time VAN telemetric network (which nowadays consists of nine measuring stations) enables the estimation of the three parameters: time (see also the methodology of natural time discussed below), epicenter and magnitude of an impending mainshock. These predictions, when the expected magnitude M is 6.0 or larger, are submitted for publication in scientific journals before the earthquake occurrence (see also below).

9.2.1 Proposal of a New Time Domain

A new concept of time, termed *natural time*, was introduced by Varotsos et al. (2001, 2002). This was followed by a sequence of papers published mainly in *Physical Review* and *Physical Review Letters*. In particular, it has been found that novel dynamical features hidden behind time series in complex systems can emerge upon analyzing them in the new time domain of *natural time*, which conforms to the desire to reduce uncertainty and extract signal information as much as possible (Abe et al. 2005). The analysis in natural time enables the study of dynamical evolution of a complex system and identifies *when the system enters a critical stage*. Hence, natural time may play a key role in predicting impending catastrophic events in general. Relevant examples of data analysis in this new time domain have been presented in a large variety of fields including *Medicine, Biology, Earth Sciences and Physics.*

For a time series comprising N events we define the natural time for the occurrence of the k-th event by $\chi_k = k/N$. Thus, we ignore the time intervals between consecutive events, but preserve their order. We also preserve their energy Q_k. We then study the evolution of the pair (χ_k, p_k), where $p_k = Q_k \Big/ \sum_{n=1}^{N} Q_k$ is the normalized energy. It has been found in earlier studies (Varotsos et al. 2001, 2002) that, in natural time analysis, the approach of a dynamical system to criticality can be identified by the variance κ_1 of natural time χ weighted for p_k, namely:

$$\kappa_1 = \sum_{k=1}^{N} p_k \chi_k^2 - \left(\sum_{k=1}^{N} p_k \chi_k \right)^2 \equiv \langle \chi^2 \rangle - \langle \chi^2 \rangle \tag{9.1}$$

Earthquakes exhibit complex correlations in time, space and magnitude, and the opinion prevails that the observed earthquake scaling laws are indicative of proximity to a *critical* point.

In the frame of natural time analysis, the quantity κ_1 serves as an order parameter of seismicity (Varotsos et al. 2005), and experiences have shown that the mainshock occurs in a few days to 1 week after the κ_1 value in the candidate epicentral area becomes approximately 0.070. This was found extremely useful in narrowing the time span of possible earthquake occurrence time prediction. However, in order to trace the time evolution of the κ_1 value, one needs to start the analysis of the seismic catalog at some time *before* mainshock(s). We chose for the starting time the initiation time of the Seismic Electric Signals (SES) activity, because SES are considered, as mentioned above, to occur when the focal zone enters the critical stage.

This procedure has been ascertained to date for several main shocks in Greece (Varotsos et al. 2005, 2006a, b, 2011) and Japan (Uyeda et al. 2009), including the prediction of the major earthquakes that occurred in Greece during 2008. For example, the occurrence time of the 6.9 earthquake on 14 February 2008, which is the strongest earthquake that occurred in Greece during almost the last 30 years, was announced as imminent on 10 February 2008 (Sarlis et al. 2008). This arouse a considerable international interest, e.g., see the two recent articles by Japanese scientists (Uyeda and Kamogawa 2008, 2010) in *EOS* Transactions of the American Geophysical Union.

9.2.2 Strains and Electric/Magnetic Fields

Point defects. As written above, there have been obtained many important results considering thermodynamics of the point defects in a regular lattice. An extension of an approach proposed by Varotsos and Alexopoulos (1986) and the related papers by Varotsos and coworkers could be consider basing on the thermodynamic functions related to the line defect density in a continuum; such approach could be based on the continuum with an additional superlattice related to the line-defects (Teisseyre and Majewski 2001). In this approach we may also include the advanced deformation processes close to fracture. We rely on the fact that some methods of the continuum mechanics can be easily applied to a material with regular lattice structure.

Thus, similarly to the role of point defects we may consider an influence related to the linear defects and their role in the appearance of electric/magnetic precursors; this approach is supported by some studies on the linear defect influence on the thermodynamical interaction processes (cf.: Teisseyre 2008; Chelidze et al. 2010).

We should explain that such considerations, related to the linear defects, do not deny or contradict the role and achievements obtained by the studies on the point defect theory and point defect thermodynamics with its important achievements in studies of the earthquake precursors.

This problem can relate to several effects: piezoelectric effect, electrokinetic and percolation processes along a fault or crack system, polarization and depolarization efects as activated by the thermal and stress release processes (dipole rotation and migration of the point defercts), bending of charged dislocations, and emissions and desorptions of the O^- radicals.

Remembering this statement, we will consider a theory describing the linear defect influence on the observed stress fields.

The direct relations between defect and electric fields relate, among other things, to:

- magnetostriction; an increase/decrease of magnetization due to applied tension;
- electrostriction; the effect independent of a reversal of electric field; $E_s \rightarrow |E_s|$;
- electro-polarization; this field is related to shear strains $\rightarrow \hat{E}_{(kl)}$;

- magneto-polarization effect;
- piezomagnetism and piezoelectricity which are the linear effects influencing the strains.

We will consider these interaction fields as the examples of such interactions in order to present the general forms for the mutual correlation between the strains, including the rotation strains, and the electric and magnetic fields.

Additionally we will consider also an effect related the appearance of the O^- radicals due to a temperature influence or some micro-fracture processes.

Thus, our aim is to present the general relations joining the stresses, especially shear and rotation ones, with the possible electric and magnetic fields including related polarizations; we include here also the role of defect distribution. Such systematical presentation might be useful in experimental pourposes and for a search for earthquake prediction methods.

We will include here also a role of the O^- current (see later on).

Magnetostriction phenomenon can be observed in ferromagnetic materials. It joins the elasticity with the magnetic and thermal fields. The related mechanism joins the extensive deformations with an external magnetic field. These phenomena have important applications of industrial purposes.

The magnetostriction relates to small material domains in which the randomly oriented rotations appear, but under an influence of the applied magnetic field there are induced some stresses. At a higher intensity of the applied magnetic field, these stresses become able to form a deformation anisotropy following orientation of the applied magnetic field; there appears a magnetoelastic coupling leading to the magnetostrictive effects, including a torque related helical anisotropy. Due to a applied magnetic field, some internal domain boundaries become shifted and a common domain rotation appears.

First, let us present some examples:

- **Magnetostriction**: a relation between strain energy density, δE^{λ}, related to some magnetic expansion domain λ, subjected to the saturated magnetization and the axial stresses; such a part of the applied energy due to an axial stress, S_{nn}, oriented along the direction of the applied magnetic field, can be presented as follows:

$$\delta E^{\lambda} = \frac{3}{2}\lambda\delta S_{nn} \tag{9.2a}$$

However, when the applied magnetic field differs by an angle, θ, from a stress orientation, then we obtain:

$$\delta E^{\lambda} = \frac{3}{2}\lambda\delta S_{nn}\sin^2\theta \tag{9.2b}$$

From this expression we may write

$$\delta E^{\lambda} = \frac{3}{2}\lambda\delta S_{nn}\sin^2\theta \propto \delta(S_{nn}E_{nn}) \propto \delta(S_{nn}S_{nn}) \to S_{nn} = Q\sin^2\theta \tag{9.2c}$$

where Q is the appropriate material constant.

There may also appear a reverse effect, that is, some magnetization due to the applied stresses.

- **Electrostriction**: for the dielectric materials we may consider the electrostriction, a property related to the random electrical domains: for an electric field applied on the opposite sides of these electric domains, they become charged in a different way and thus mutually correlated; electrostriction means the electro-mechanical interactions in crystalline materials with a counterpart of stresses, S_{ik}:

$$E_{ki} = s_{kins}S_{ns} + d_{kip}E_p \text{ and } D_n = n_{nik}S_{ik} + \varepsilon_{ns}E_s \tag{9.3a}$$

where $E_k$is the electric field and $D_i = \varepsilon_{ik}E_k$ is the electric displacement (charge density displacement), and ε is permittivity.

There may appear also a mutual interaction between the induced strains, δE_{lp}, and the polarizations, Π_k:

$$\delta E_{lp} = \delta Q_{lpik}\varepsilon_{iks}\varepsilon_{smn}\Pi_m\Pi_n \tag{9.3b}$$

Thus, formally we might deduce the polarization contributions, Π, to the strains (or stresses) as proportional to their rotation product; thus, an expression for related stress may be presented with the adequate constants, Q_{lpik}, as:

$$S_{lp} = Q_{lpik}\varepsilon_{iks}\varepsilon_{smn}\Pi_m\Pi_n \tag{9.3c}$$

Electrostriction and magnetostriction are related to a property of all electrical non-conductors, or dielectrics, that change their shape under the application of the electric or magnetic fields. These examples give us an idea how to construct some general relations joining the electric and magnetic interactions with stresses; we include here also the polarization fields.

We may present also the general formulae presenting the influences of the non-mechanical fields (e.g., electric, magnetic fields and polarization ones) on the stresses and strains (e.g., Chelidze et al. 2010). We may define the stresses or strains in relation to some non-mechanical fields, G (electric, magnetic and polarization related); we limit ourselves to a range of linearity:

$$S_{kl} = \lambda\delta_{ij}\sum_s E_{ss} + \delta_{ij}G + 2\mu E_{ij} + G_{ij} \tag{9.4a}$$

or

$$\begin{aligned} S_{(kl)} &= \lambda\delta_{kl}\sum_s E_{ss} + \delta_{kl}G + 2\mu E_{(kl)} + \delta_{kl}G + G_{(kl)}, \\ S_{[kl]} &= 2\mu E_{[kl]} + \varepsilon_{kls}G_s + G_{[kl]} \end{aligned} \tag{9.4b}$$

Here we have presented the possible influences of the non-mechanical fields, $\mathbf{G} = \left\{G\delta_{ij}, G_{ij}, G_{(kl)}, \varepsilon_{kls}G_s, G_{[kl]}\right\}$.

For the specific cases we may distinguish these influences using similar symbols with the upper indexes, $\mathrm{G} = \{\mathrm{G}^M, \mathrm{G}^\Pi, \mathrm{G}^{\Pi g}\mathrm{G}^E, \mathrm{G}^M, \mathrm{G}^\Pi, \mathrm{G}^{\Pi g}, \ldots\}$, for the influences of electric (E_k), magnetic (B_s), polarization (Π_s), and polarization gradient ($\mathrm{G}^{\Pi g}$) effects; e.g., we may use some of the following symbols:

$$\begin{aligned}
&\left\{\delta_{kl}G^E|E_s|,\ G_k^E E_l,\ \varepsilon_{kls}\sum_s G_s^E E_s,\ \sum_s G_{s(kl)}^E E_s,\ \sum_s G_{s[kl]}^E E_s\right\}\\
&\left\{\delta_{kl}G^M|B_s|,\ G_k^M B_l,\ \varepsilon_{kls}\sum_s G_s^M B_s,\ \sum_s G_{s(kl)}^M B_s,\ \sum_s G_{s[kl]}^M B_s\right\}\\
&\left\{\delta_{kl}G^\Pi|\Pi_s|,\ G_k^\Pi \Pi_l,\ \varepsilon_{kls}\sum_s G_s^\Pi \Pi_s,\ \sum_s G_{s(kl)}^\Pi \Pi_s,\ \sum_s G_{s[kl]}^\Pi \Pi_s\right\}\\
&\left\{\delta_{kl}G^{\Pi^g}|\Pi_{ss}|,\ G^{\Pi^g}\Pi_{kl}^g,\ \varepsilon_{kls}\sum_s G_s^{\Pi^g}\Pi_s^g,\ \sum_s G_{s(kl)}^{\Pi^g}\Pi_s^g,\ \sum_s G_{s[kl]}^{\Pi^g}\Pi_s^g\right\}
\end{aligned} \tag{9.4c}$$

The non-mechanical fields considered here, E_k, B_k, Π_k, Π_k^g,may influence the stresses; e.g., we can write for an electrostriction dependent only on magnitude of electric field:

$$S_{(kk)} = \lambda\sum_s E_{ss} + G^E|E_k| + 2\mu E_{(kk)} \tag{9.4d}$$

It is worth to remember here that the independent strain fields, axial, deviatoric and rotational strains, $\hat{E}_{(lk)}$, $E_{[lk]}$, can be released at the source due to a common process, but with a possible phase shift, e.g., $E_{(kl)}^D, \pm\, \mathrm{i}E_{[kl]}$.

Further on, we can also formally write the direct relations between the defects and electric, magnetic and polarization fields (Chap. 5, Eq. 5.3):

$$\begin{aligned}
&\alpha_{pl} - \frac{1}{2}\delta_{pl}\sum_s \alpha_{ss} = \varepsilon_{pmk}\frac{\partial}{\partial x_m}\left[E_{(kl)} + \frac{1}{2\mu}\left(\delta_{kl}G + \varepsilon_{kls}G_s + G_{(kl)}\right)\right] +\\
&\varepsilon_{pmk}\frac{\partial}{\partial x_m}\left[E_{[kl]} + \frac{1}{2\mu}\left(\varepsilon_{kls}G_s + G_{[kl]}\right)\right] - \frac{v}{1+v}\delta_{kl}\left(\sum_s E_{ss} + \frac{1}{2\mu}\left(3G + \sum_s G_{ss}\right)\right)
\end{aligned} \tag{9.5a}$$

and

$$\theta_{pq} = \varepsilon_{pmk}\frac{\partial}{\partial x_m}\left(E_{[kl]} + \varepsilon_{kqs}G_s + G_{[kl]}\right) \tag{9.5b}$$

These differential equations estimate directly an influence of the non-mechanical fields on defects. Here we have assumed that these processes take place under thermal equilibrium.

Now, we may consider some particular cases.

- **Piezoelectric effects.** The piezoelectric effect can be presented by the following relations (after Toupin 1956; see: Mindlin 1972; Teisseyre 2001a, b) with $G_{kl} = -\sum_n e_{nkl} E_n$ (cf., Eq. 9.2a):

$$S_{kl} = \lambda\delta_{ij}E_{ss} + \delta_{ij}G + 2\mu E_{kl} - \sum_n e_{nkl}E_n \rightarrow$$
$$S_{(ij)} = \lambda\delta_{ij}E_{ss} + \delta_{ij}G + 2\mu E_{(ij)} - \sum_n e_{n(ij)}E_n, \quad (9.6a)$$
$$S_{[ij]} = 2\mu E_{[ij]} - \sum_n e_{n[ij]}E_n$$

where E_k is the electric field, e_{kij} represents the piezoelectric parameters.

The piezoelectric parameters, e_{kij}, can be separated into the symmetric and antisymmetric parts:

$$e_{kij} = -2\mu\left(f_k\delta_{ij} + \varepsilon_{kij}g\right) \text{ and } F\delta_{ij} = f_k\delta_{ij}E_k, G_{[ij]} = \varepsilon_{kij}gE_k \quad (9.6b)$$

where $\varepsilon_{kij} = \{1, -1, 0\}$ is the fully asymmetric tensor.

The equivalent relation between the defect density and this piezo-electric field becomes:

$$\alpha_{pl} - \frac{1}{2}\delta_{pl}\alpha_{ss} = e^0\varepsilon_{pmk}\frac{\partial}{\partial x_m}\left(E_{kl} + \delta_{kl}G - \frac{1}{2\mu}\sum_n e_{nkl}E_n\right) \quad (9.7a)$$

or

$$\alpha_{pl} - \frac{1}{2}\delta_{pl}\alpha_{ss} = \varepsilon_{pmk}\frac{\partial}{\partial x_m}\left(\left(E_{(kl)} + f_s\delta_{kl}E_s\right) - \frac{\nu}{1+\nu}\delta_{kl}\left(E_{ss} + 3f_sE_s\right) + E_{[kl]} + \varepsilon_{skl}gE_s\right)$$
$$\theta_{pq} = \varepsilon_{pmk}\frac{\partial\chi_{kq}}{\partial x_m} = \frac{1}{l}\varepsilon_{pmk}\frac{\partial}{\partial x_m}\left(E_{[kq]} + \varepsilon_{skq}gE_s\right) \quad (9.7b)$$

For the piezo-electric constants in relation to the different crystallographic classes we may refer to Nowacki (1986).

- **Polarization gradient theory**. After Mindlin (1972), the internal energy depends also on the polarization gradient:

$$\Pi_{ij} = \frac{\partial\Pi_i}{\partial x_j} \quad (9.8)$$

where polarization, $\Pi_i = D_i - \varepsilon E_i$, is defined by difference of the electric displacement, D, and electric field, E, with ε being the permittivity of vacuum.

The gradient theory, related to an electric polarization, follows from the fact that, under the applied load, the displacements of a moving dislocation core (electricaly charged) influence the surrounding defect cloud (such a cloud should have an opposite charge, compensating that of the dislocation core).

For this polarization problem, the constitutive relations (Mindlin 1972; Nowacki 1986) with the respective material constants can be written as follows:

$$\begin{aligned} S_{ij} &= \lambda\delta_{ij}\sum_{s} E_{ss} + \delta_{ij}G + 2\mu E_{ij} + G_{ij} + = \lambda\delta_{ij}\sum_{s} E_{ss} + \delta_{ij}G \\ &= 2\mu E_{ij} + f_{kij}\Pi_k + d_{klij}\Pi_{kl} \end{aligned} \tag{9.9a}$$

According to this relation, and with $G = 0$, we can present these relations for the asymmetric stresses in the following form (cf., Teisseyre 2001a, b):

$$\begin{aligned} S_{(ij)} &= \lambda\delta_{ij}\sum_{s} E_{ss} + 2\mu E_{(ij)} + f_{k(ij)}\Pi_k + d_{kl(ij)}\Pi_{kl}\,, \\ S_{[ij]} &= 2\mu E_{[ij]} + f_{k[ij]}\Pi_k + d_{kl[ij]}\Pi_{kl} \end{aligned} \tag{9.9b}$$

For a direct relation with the defects we can write (cf., Chap. 5):

$$\begin{aligned} &\alpha_{pl} - \frac{1}{2}\delta_{pl}\alpha_{ss} = \varepsilon_{pmk}\frac{\partial}{\partial x_m}\left[E_{(kl)} + \frac{1}{2\mu}\left(\sum_{s} f_{s(kl)}\Pi_s + \sum_{i,j} d_{ij(kl)}\Pi_{ij}\right)\right] + \\ &\varepsilon_{pmk}\frac{\partial}{\partial x_m}\left[E_{[kl]} + \frac{1}{2\mu}\left(\sum_{s} f_{s[kl]}\Pi_s + \sum_{i,j} d_{ij[kl]}\Pi_{ij}\right)\right] - \\ &\frac{\nu}{1+\nu}\varepsilon_{pmk}\frac{\partial}{\partial x_m}\frac{\delta_{kl}}{2\mu}\left(\sum_{s,k} f_{s(kk)}\Pi_s + \sum_{i,j,k} d_{ij(kk)}\Pi_{ij}\right) \end{aligned} \tag{9.10a}$$

and

$$\theta_{pq} = \varepsilon_{pmk}\frac{\partial}{\partial x_m}\left[E_{[kl]} + \frac{1}{2\mu}\left(\sum_{s} f_{s[kl]}\Pi_s + \sum_{i,j} d_{ij[kl]}\Pi_{ij}\right)\right] \tag{9.10b}$$

Moreover, we shall note that some experiments (see e.g., Hadijcondis and Mavromatou 1994, 1995) indicate that the anomalous piezoelectric effects observed in the laboratory experiments correspond to the time rate of the applied load.

An array of the vortex-defects can help us to approximate the fragmentation cracks (similarly as an array of dislocations approximates a crack).

The problem of magnetostrictive effects can be treated in a similar way.

In the next Chap. 10 we will consider the fracture probability relations under an applied external load consisting of a constant pressure and the additional shear or/and rotation vibrating stresses in a wide frequency range.

We should include the additional influences related to the electric and magnetostrictive effects, first of all those related to polarization, piezoelectricity and polarization gradient; these problems could be treated in a similar way.

Separately, we may consider the thermal interaction. For a thermal influence on the asymmetric strains, we may write a generalized relation:

$$S_{(kl)} = \lambda \delta_{ij} \sum_s E_{ss} + \lambda \delta_{kl} \alpha^{\text{th}} (T - T_0) + 2\mu E_{(kl)}, \quad S_{[kl]} = 2\mu E_{[kl]} \tag{9.11a}$$

and for defects we obtain from Eq. (9.8):

$$\begin{gathered} \alpha_{pl} - \frac{1}{2}\delta_{pl}\alpha_{ss} = \\ \varepsilon_{pmk}\frac{\partial}{\partial x_m}\left(E_{(kl)} + \frac{1}{2\mu}\sum_s f_{s(kl)}\Pi_s + \frac{1}{2\mu}\sum_{i,j} d_{ij(kl)}\Pi_{ij} + E_{[kl]} + \varepsilon_{kls}G_s + G_{[kl]}\right) - \\ \frac{\nu}{1+\nu}\varepsilon_{pml}\frac{\partial}{\partial x_m}\left(\sum_s E_{ss} + \alpha^{\text{th}}(T - T_0) + \frac{1}{2\mu}\sum_{s,k} f_{s(kk)}\Pi_s + \frac{1}{2\mu}\sum_{i,j,k} d_{ij(kk)}\Pi_{ij}\right) \end{gathered} \tag{9.11b}$$

A relation between the thermal field and the dislocations becomes:

$$\alpha_{pl}^{\text{edge}} = -\alpha^{\text{th}} \frac{\nu}{1+\nu} \varepsilon_{pml} \frac{\partial}{\partial x_m}(T - T_0) \tag{9.12}$$

and there is no such contribution related to the screw dislocations.

Finally, we should remark that the influences of the external fields, introduced above, might be emitted from a source independently. Therefore, we can insert to any of these independent contributions the additional factors, e.g., for the fields,$S_{(kl)}$, $S_{[kl]}$, or for other fields, the phase factors, e.g., ς^0 and χ^0, equal to $\{\varsigma^0, \chi^0\} = \{0, \pm 1, \pm \mathrm{i}\}$, that means that the possible phase shifts are as follows: 0, $\pm\pi/2$ or $\pm\pi$.

Thus, instead of relations (9.4) we write:

$$\begin{aligned} S_{(kl)} &= \lambda \delta_{kl} \sum_s E_{ss} + \varsigma^{01} G + 2\mu E_{(kl)} + \varsigma^{02} \delta_{kl} G + \varsigma^{03} \chi^{01} G_{(kl)}, \\ S_{[kl]} &= 2\mu E_{[kl]} + \chi^{01} \varepsilon_{kls} G_s + \chi^{02} G_{[kl]} \end{aligned} \tag{9.13}$$

Thus, the considered shear and rotation fields, and electric and magnetic influences, can be observed as the mutually phase shifted elements.

Finally, we may consider the interaction chains; e.g., electric and acoustic effects; it seems reasonable to believe that coupling between the mechanical and electric (or electric polarization) field proceeds in an instantaneous manner, because such effects follow from the displacement of the ions. However, as mentioned above, in general the interaction between the mechanical fields can proceed with the phase delay due to the release-rebound sequence. Hence, we may have the different interaction chains like, e.g., the following ones:

$$E_{(nl)}^D \to \mathrm{i}E_{[nl]} \to \mathrm{i}\Pi_s \tag{9.14}$$

where the shears coupled to the phase delayed rotations lead to polarization effects.

Also a pressure variation, p, may initiate, as a mechanical forcing, some chain related to electric, E_s, and mechanical, E_{nl}, fields:

$$p \to E_s \to E_{[nl]} \to \mathrm{i}E_{(nl)} \tag{9.15}$$

where here above we have added to the electric field variations also strain rotation effects and some shear strain releases revealed by the acoustic bursts occurring with the phase delay.

Such different chains have been observed in the experiments on the different synchronizations between the acoustic bursts and electric load related to additional oscillations (Chelidze et al. 2010).

We should take into account that any crushing process relates to the shear strength; even under the confining pressure a crushing process relates to shears—real or induced; the effective energy of the induced crushing processes relates to the resonance between an internal process and the given applied mechanical load or other applied field. There remain several problems related to a kind of additional frequency dependent load (e.g., electric or other fields). The grain fracture energy, E^{GF}, can be related to the circular defects; we consider an the upper mantle solid material under the pressure with the vertical gradient, hence we can use the cylindrical coordinates. We might assume that there are three kinds of defects as related to the defect radius: the smallest ones relate to the real point defects with radius given as the Plank length (ca 10^{-34} m, the point defects), while two other defects can be related to the internal grain dimensions (two kinds of grain are frequently considered in problems of material fracture).

Interaction processes. The axial field shall be considered always separately, while the shear deviatoric and rotation fields, $E_{lk} \to \hat{E}_{(lk)}$ and $E_{lk} \to \tilde{E}_{[lk]}$, can be released at the source as coupled fields shifted in phase:

$$E^D_{kl} \leftrightarrow \pm\mathrm{i}\tilde{E}_{[lk]} \tag{9.16a}$$

For the interaction processes with the non-mechanical interaction stress parts we have written a very general form of the constitutive laws (9.1); the magneto-strictive effects could be treated in a similarl way.

Of course, we may note that, from the other side, we should add the reverse relations presenting the mechanical influences on the considered electro-magnetic fields.

Finally, we should mention that the considered shear and rotation fields, and electric and magnetic influences, could be observed as the mutually phase shifted elements. The release-rebound theory would not only apply to the shear and rotations, $\omega_{lk} \leftrightarrow \pm\mathrm{i}\hat{E}_{(lk)}$, but also to the interaction fields; we might observe the phase shifted interaction coupling, like a chain, e.g.:

$$\hat{E}_{(nl)} \to \pm\mathrm{i}\tilde{E}_{[nl]} \to \pm\mathrm{i}\Pi_s \tag{9.16b}$$

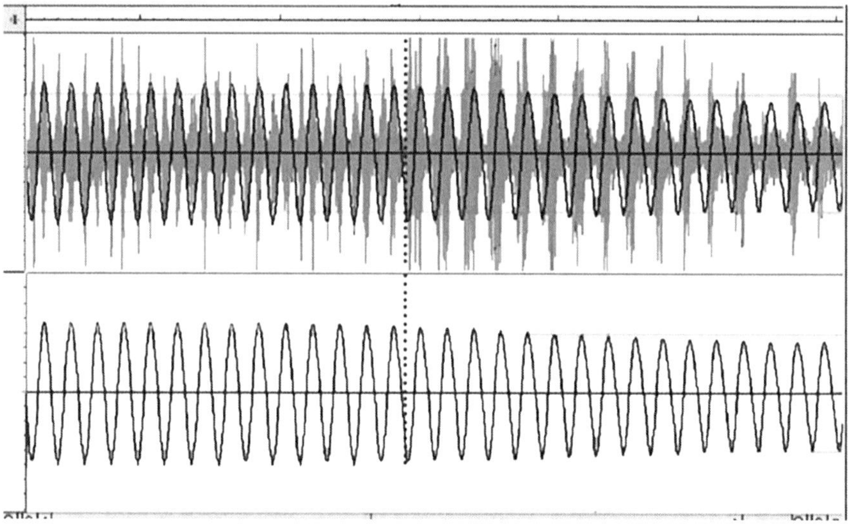

Fig. 9.1 The acoustic bursts and the electric load oscillation as the initial field (modified after Chelidze et al. 2010). The system includes also a constant pressure load and a constant electric voltage. In the *left part*, a constant electric load is 400 V and on the *right part* it is 1900 V; the applied oscilating voltage part has 800 V amplitudes. On a *left side*, the acoustic bursts (marked by the shadow oscillations) appear with a double frequency, while under higher constant voltage (*right side*) the frequency of acoustic bursts is equal to the applied electric frequency; the bursts appear more rarely, that is, in accordance with maxima of oscillating voltage, and due to the applied constant voltage (a small phase shift appears)

We would like to mention here the strain release processes revealed by the acoustic bursts occurring with the phase delay to the electric load oscillations (see: Chelidze et al. 2010; Teisseyre 2010); Fig. 9.1 presents the electric load oscillations and the acoustic bursts.

It seems that our considerations might be useful to estimate, from a formal point of view, a number of possible interactive phenomena as presented by the relations (9.4a, b, c) for

$$\mathrm{G} = \left\{G\delta_{ij}, G_{ij}, G_{(kl)}, \varepsilon_{kls}G_s, G_{[kl]}\right\} \text{ and } \mathrm{G} = \left\{\mathrm{G}^M, \mathrm{G}^\Pi, \mathrm{G}^{\Pi g}\mathrm{G}^E, \mathrm{G}^M, \mathrm{G}^\Pi, \mathrm{G}^{\Pi g}, \ldots\right\} \tag{9.17}$$

However, this is not exactly so; nature remains still more complicated, as will be shown in the last example.

The O^- current. Finally we should add here an important effect related to the O^- current. Acording to Freund et al. (1994), the properties of the O^- radical are "fuzzy"; this radical presents the electron defect, chemically strongly active, and being a paramagnetic ion.

In several papers Freund and his co-workers studied the release processes of such radicals, especially those released from the defect sites in rocks and in some

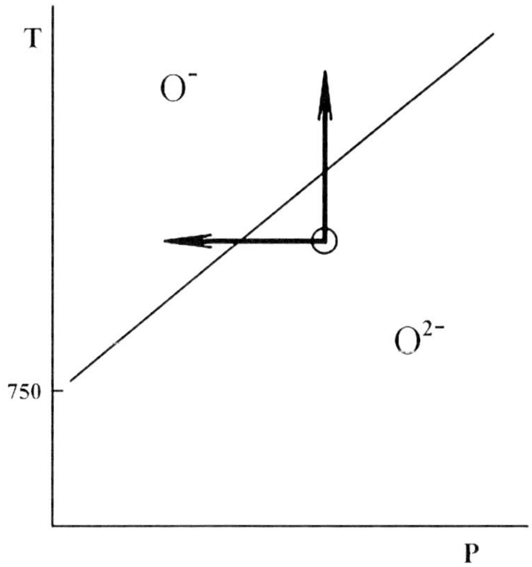

Fig. 9.2 Plot of temperature (°C) vs. pressure T. The domain of possible appearance of the O^- radicals in MgO is in the upper *left part*, and the domain in which such a process is not expected is in the lower *right part* (after Freund et al. 1994, modified)

oxides. Such studies become very important for the generation mechanism of electric field and also for current generated in the seismic sources.

A release of the O^- radicals relates to the thermally activated anomalies or some stress release (Freund et al. 1994, Malinowski 2001), like a micro-fracture.

Finally, we should add to these studies also the release of the O^- radicals, as the thermally activated anomalies (Freund et al. 1994; Malinowski 2001); in MgO at temperatures 200 – 300 °C there appear the two paramagnetic ions O^- from O^{2-}; Fig. 9.2 presents a domain in which the O^- radicals can appear.

We may point out a possible involvement of these effects in the formation of earthquake precursors; the related charge diffusion towards a body surface forms an electric current as a hole conductivity (see also Teisseyre 2001a, b).

Thus, when due to some micro-fractures a micro-part, O^-, appears as separated from a chemical compound, then the "O^-" current may appear.

This current spreads almost without any mechanical processes; the current progress of this electric field towards the Earth surface is based on an electron jump from some particle containing an element, O^{2-}, towards a micro-part, O^-; in this way a negative electric current moves up (Fig. 9.3).

Considering the effects caused by this special electric current related to the O^- radicals, the hole electric current, we may remark that the related precursory

$$\begin{array}{ccc} O^2 & O^- & O^- \\ \text{electron falls down: } \downarrow \text{ leading to: } | & \text{, forming a current progress up: } \uparrow \\ O^- & O^2 & O^- \end{array}$$

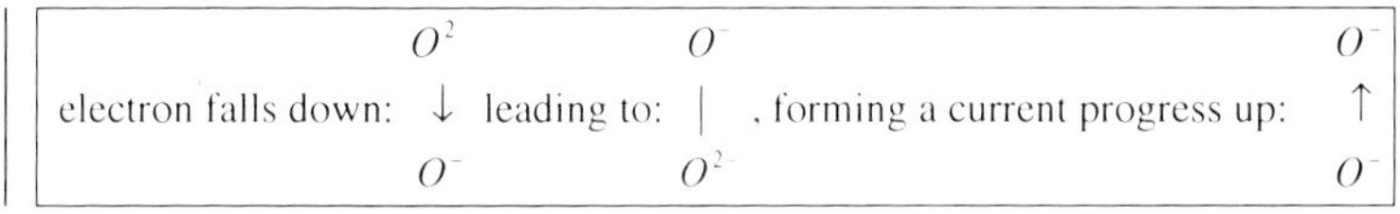

Fig. 9.3 Propagation mechanism of the electric current towards the Earth surface, as related to the O^- radicals

signals may be not detected by any mechanical system; practically, only the electric effects will remain to be detected, probably even at remote sites. We may also note that a charge release process and charge diffusion towards a body surface along some plane (e.g., tectonic) could cause an increase of the electric conductivity along that plane.

We should remark that this effect, related to the O^- radicals, might be explained according to our formalism including the molecular displacements (cf. Chap. 2, Eqs. 2.11a, b); an electron from O^{2-}(in MgO) could drop down causing the position of a O^- radical to be shifted up and, in this way, a molecular wave propagation towards the Earth surface becomes realized.

References

Abe S, Sarlis NV, Skordas ES, Tanaka HK, Varotsos PA (2005) Origin of the usefulness of the natural time representation of complex time series. Phys Rev Lett 94:170601

Chelidze T, Matcharashvili T, Lursmanashvili O, Varamashvili N, Zhukova E, Meparidze E (2010) Triggering and synchronization of stick-sip: experiments on spring-slider system. In: De Rubeis V, Czechowski Zb, Teisseyre R (eds) Synchronization and triggering: from fracture to earthquake processes. Springer-Verlag, Berlin Heidelberg, p 123–164

Freund F, Whang EJ, Lee J (1994) Highly mobile charge carriers in minerals. Electromagnetic phenomena related to earthquake prediction. Terra Science Publication Chem, Tokyo, pp 271–277

Hadijcondis V, Mavromatou C (1994) Transient electric signals prior to rock failure under uniaxial compression. Geophys Res Lett 21:1687–1690

Hadijcondis V, Mavromatou C (1995) Electric signals during uniaxial compression of rock samples: their correlation with preseismic electic signals. Acta Geophys Pol 41(1):49–61

Malinowski S (2001) Physical and chemical properties related to defect structure of oxides and silicates doped with water and carbon dioxide. In: Teisseyre R, Majewski E (eds) Earthquake thermodynamics and phase transformations in the Earth's interior. (International geophysics series vol 76). Academic Press, San Diego, pp 441–459

Mindlin RD (1972) Elasticity, piezoelectricity and crystal lattice dynamics. J Elast 2:217–282

Nowacki W (1986) Theory of Asymmetric Elasticity, PWN-Pergamon Press, Warszawa-Oxford.

Sarlis NV, Skordas ES, Lazaridou MS, Varotsos PA (2008) Investigation of seismicity after the initiation of a seismic electric signal activity until the main shock. Proc Japan Acad Ser B 84:331–343

Teisseyre R (2001a) Shear band thermodynamic model of fracturing. In: Teisseyre R, Majewski E (eds) Earthquake thermodynamics and phase transformations in the Earth's interior. (International geophysics series vol 76). Academic Press, San Diego

Teisseyre R (2001b) Evolution, propagation and diffusion of dislocation fields. In: Teisseyre R, Majewski E (eds) Earthquake thermodynamics and phase transformations in the Earth's interior. (International geophysical series vol 76). Academic Press, San Diego, pp 167–198

Teisseyre R (2008) Asymmetric continuum: standard theory. In: Teisseyre R, Nagahama H, Majewski E (eds) Physics of asymmetric continua: extreme and fracture processes. Springer, Berlin, pp 95–109

Teisseyre R (2010) Fluid theory with asymmetric molecular stresses: difference between vorticity and spin equations. Acta Geophys 58(6):1056–1071

Teisseyre R, Majewski E (2001) Thermodynamics of line defects and earthquake thermodynamics. In: Teisseyre R, Majewski E (eds) Earthquake thermodynamics and phase transformations in the Earth's interior. (International geophysics series vol 76). Academic Press, San Diego, pp 261–278

Toupin RA (1956) The elastic dielectrics. J Rat Mech Anal 5:849–915

Uyeda S, Kamogawa M (2008) The prediction of two large earthquakes in Greece. EOS 89(39):363

Uyeda S, Kamogawa M (2010) Reply to comment on 'The prediction of two large earthquakes in Greece'. EOS 91(18):162

Uyeda S, Kamogawa M, Tanaka H (2009) Analysis of electrical activity and seismicity in the natural time domain for the volcanic-seismic swarm activity in 2000 in the Izu Island region. Jpn J Geophys Res 114:B02310

Varotsos P (2005) The physics of seismic electric signals. TerraPub, Tokyo, p 338

Varotsos P, Alexopoulos K (1984a) Physical properties of the variations of the electric field of the earth preceding earthquakes, I. Tectonophysics 110:73–98

Varotsos P, Alexopoulos K (1984b) Physical properties of the variations of the electric field of the earth preceding earthquakes, II. Determination of epicenter and magnitude. Tectonophysics 110:99–125

Varotsos PA, Alexopoulos KD (1986) Thermodynamics of point defects and their relation with bulk properties. North-Holland, Amsterdam, p 474

Varotsos P, Lazaridou M (1991) Latest aspects of earthquake prediction in Greece based on seismic electric signals I. Tectonophysics 188:321–347

Varotsos P, Alexopoulos K, Lazaridou M (1993) Latest aspects of earthquake prediction in Greece based on seismic electric signals II. Tectonophysics 224:1–37

Varotsos P, Sarlis N, Skordas E (2001) Spatiotemporal complexity aspects on the interrelation between seismic electric signals and seismicity. Practica Athens Acad 76:294–321

Varotsos PA, Sarlis NV, Skordas ES (2002) Long-range correlations in the electric signals that precede rupture. Phys Rev E 66:011902

Varotsos PA, Sarlis NV, Skordas ES (2003) Electric fields that "arrive" before the time-derivative of the magnetic field prior to major earthquakes. Phys Rev Lett 91:148501

Varotsos PA, Sarlis NV, Tanaka HK, Skordas ES (2005) Similarity of fluctuations in correlated systems: the case of seismicity. Phys Rev E 72(8):041103

Varotsos PA, Sarlis NV, Skordas ES, Tanaka HK, Lazaridou MS (2006a) Entropy of seismic electric signals: analysis in natural time under time reversal. Phys Rev E 73:031114

Varotsos PA, Sarlis NV, Skordas ES, Tanaka HK, Lazaridou MS (2006b) Attempt to distinguish long-range temporal correlations from the statistics of the increments by natural time analysis. Phys Rev E 74:021123

Varotsos PA, Sarlis NV, Skordas ES (2011) Natural time analysis: the new view of time. Precursory seismic electric signals, earthquakes and other complex time-series. Springer-Verlag, Berlin

Chapter 10
Quantum Analogies

10.1 Introduction

We consider here the fracture processes in a body under different frequency loads and a possibility of an existence of some resonance frequencies. In a reverse sense, the problem of resonance frequency can be related to the classical question of the black body radiation, that is, the radiation in thermodynamic equilibrium—a question related to the first fundamental problem leading to the quantum mechanics. We are trying to use some similar considerations in relation to the fracture processes, e.g., the earthquake events, under the thermodynamic equilibrium.

10.2 Quantum Analogies

We consider here the fracture processes in a body under different frequency loads and define a possible resonance frequency. In a reverse sense, the problem of resonance frequency may be related to the classical question of the black body radiation, that is, the radiation in thermodynamic equilibrium. This question was a first fundamental problem related to the birth of quantum mechanics.

Here and further on we omitt the references related to the elements of quantum theory; in our considerations we use only the basic elements of quantum theory which could be found in anyelementary textbook on these topics.

Here, when considering the fracture processes, e.g., during an earthquake event, we will consider the fracture processes under thermodynamic equilibrium.

First, let us recollect the black body radiation in quantum mechanics: we write the Planck relation describing the energy radiated per unit wavelength:

$$e(\lambda) = \frac{8\pi vh}{\lambda^4}\left(\exp\left(\frac{vh}{KT}\right) - 1\right)^{-1} \tag{10.1}$$

where λ is the wavelength, h the Planck constant, K the Boltzmann constant, T the temperature, and assume that the product, λT, remains constant.

R. Teisseyre and M. Teisseyre-Jeleńska, *Asymmetric Continuum*,
GeoPlanet: Earth and Planetary Sciences, DOI: 10.1007/978-3-642-31860-3_10,

We remind the basic relations, e.g.:

$$e = h\nu = mc^2, \quad 1/\nu = \tau, \quad \lambda/c = \tau, \quad p_i = \frac{h}{\lambda} n_i \tag{10.2}$$

where p_i is the moment of momentum.

However, in an equilibrium thermodynamics (black body radiation) the energy, $e = h\nu$ (Eq. 10.2), becomes modified as presented in Eq. (10.1) and will reach a maximum.

However, the proposed analogy related to the fracture processes would mean, in a reverse way, a minimum of energy required for the fracture in a thermodynamic equilibrium. Such a fracture process can be related to the molecular transport at fracture recess; see Chap. 3 and Eq. 3.23. Moreover, we assume that such a fracture relates to the oscillating shear load at the different frequencies and to a constant part of shears or constant pressure load. We postulate, in analogy to the relativistic formula, that a crushing energy is, on the one side, proportional to frequency and on the other side, to the mass of grains forming a body structure.

The individual fracture processes start and remain at the quantum level; therefore, for the microscopic fracture events, e.g., the individual and collective events, we introduce, besides the macroscopic continuum relations, also some other relations similar to the quantum theory. Moreover, for some associated electro-magnetic phenomena we may include some relativistic quantum-like relations, e.g., the Klein–Gordon equation. This proposed approach will be called the quantum-like state. In that case, the term "black body radiation" shall mean the seismic wave radiation being in thermodynamic equilibrium within a given material recess related to the seismic source domain.

We assume that fracture processes and the emission of elastic waves run in the thermodynamic equilibrium within the related fracture recess. We assume that a shear load is given by a sum of the constant part and an additional oscillating part having a given frequency. Moreover, the constant pressure part and that of the moment of momentum can be included.

To this end, we should introduce some relations which could lead us to such a quantum-like approach. Thus, we may assume the following relations for the energy needed for fracture of a grain element; a released energy unit in the process of grain destruction may be given as:

$$\Delta e = \frac{\Delta \rho}{c^2} \frac{4\pi}{3} r_0^3, \quad \Delta m = \rho \frac{4\pi}{3} r_0^3, \tag{10.3}$$

where $\Delta \rho$ means an equivalent decrease of density due to a crushing process inside a given volume element and r_0 represents an equivalent radius of the released mass, Δm, from a compact body.

The released binding energy, at an applied break frequency, ν, might be presented in analogy to the relativistic formula (Eq. 10.1). According to this Planck relation we write a similar expression for the applied radiated energy (due to an oscillating load part) at a thermal equilibrium inside a fracture domain (recess, cavity) and per unit wavelength:

$$E^F(\lambda) = \frac{8\pi\nu H}{\lambda^4}\left(\exp\left(\frac{\nu H}{KT}\right) - 1\right)^{-1} \tag{10.4}$$

where according to the Planck relation (Eq. 10.1), the fracture energy is given at a thermal equilibrium inside a fracture domain (recess, cavity) and per unit wave length λ (the product λT should be constant), K is the Boltzmann constant, T the temperature, and H is a characteristic material crushing constant (introduced here in analogy to the Planck constant).

A characteristic material crushing constant, H, may have, of course, different values depending on a material kind, temperature and other factors, like static part of a stress load. Taking these circumstances into account we might present an energy expression needed for a crushing process of a grain with radius r_0 related to the equivalent released mass, Δm, from a compact body at a constant temperature:

$$E^F = H\nu = \Delta\rho\, c^2 \frac{4\pi}{3} r_0^3 = \rho\, V_S^2 \frac{4\pi}{3} r_0^3 \tag{10.5a}$$

This value of the applied released energy should be proportional to the applied frequency and, on the other hand, equal to energy of the mass element released by a fracture at a constant temperature (Eq. 10.3).

Moreover, $\Delta\rho$ arrive from (10.5a) to an definition of the density drop, $\Delta\rho$, related to the density change in a related volume:

$$E^F = H\nu = \rho V_S^2 \frac{4\pi}{3} r_0^3, \quad \frac{\Delta\rho}{\rho} = \frac{V_S^2}{c^2} \tag{10.5b}$$

where V_S means the S-wave velocity, r_0 is the radius of the crushed grain element, $\Delta\rho$ means here density decrease in that grain due to a fracture process, H is material constant, and $\nu = 1/\tau$ means frequency.

However, the fracture crushing energy cannot grow to infinity with growing frequency; we should rather expect some frequency maximum (or maxima for a body with different grains) and a further decrease of released energy at greater frequencies. In this way we approach the basic quantum-like relation, Eq. (10.4).

The relation (10.4) means that at a given value of the constant H, the crushing susceptibility will reach the greatest value at $\nu = \nu_0$, defined as a maximum frequency related to the highest aptitude for fracture. Such a maximum for fracture susceptibility means a minimum of a material strength: a fracture under constant load (shear or pressure) and due to additional vibrating shears and rotations could be schematically presented as in Fig. 10.1.

We repeat that the radiated energy per unit wavelength may lead us to a similar relation for the released energy in fracture process and described by the Planck-like relation:

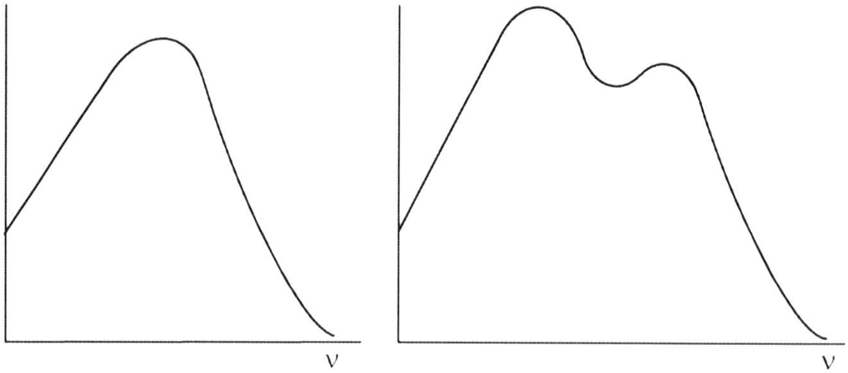

Fig. 10.1 A schematic view of possible resonance frequencies, at thermodynamic equilibrium, for the minima of crushing resistances (the vertical axis relates to the inverse of crushing resistances). The *right* diagram presents a case with two resonance frequencies

$$e(\lambda) = \frac{8\pi ch}{\lambda^5}\left(\exp\left(\frac{\nu h}{KT}\right) - 1\right)^{-1} \rightarrow e(\lambda) = \frac{8\pi VH}{\lambda^5}\left(\exp\left(\frac{\nu H}{KT}\right) - 1\right)^{-1} \quad (10.5c)$$

And in such way we can obtain the following equation for the resonance frequency at an extremum, V_E:

$$\begin{aligned} e^{\nu} &= \frac{8\pi H\nu^3}{V^3}\left(\exp\left(\frac{H\nu}{KT}\right) - 1\right)^{-1} \rightarrow \\ \partial e^{\nu}/\partial\nu &= \frac{24\pi H\nu^2}{V^3}\left(\exp\left(\frac{H\nu}{KT}\right) - 1\right)^{-1} - \frac{8\pi H\nu^3}{V^3}\frac{H}{KT}\exp\left(\frac{H\nu}{KT}\right)\left(\exp\left(\frac{H\nu}{KT}\right) - 1\right)^{-2} = 0 \\ \nu_E &= \frac{3KT}{H}\frac{\exp\left(\frac{H\nu}{KT}\right)}{\exp\left(\frac{H\nu}{KT}\right) - 1} \end{aligned} \quad (10.5d)$$

A constant part of shears and pressure (below the static strength) acting on a body could lead us to the different values of strength related to the vibrating shear, or rotation, loads; from this point of view we may treat such bodies under the given constant pressure and shear as different species.

We may remind (cf., Chap. 5) that any pressure load, even constant, influences the defect distributions, and the defect arrays could be formed; in consequence, there appear the induced shear and rotation strains (Teisseyre and Gorski 2012).

However, the fracture crushing energy cannot grow to infinity with growing frequency; we expect its maximum (or maxima for a body with different grains) and its further decrease for greater frequencies. This means that at given value of the constant H, the crushing ability will reach the greatest value at $\nu = \nu_0$, defined as frequency related to the highest aptitude for fracture. Such a maximum for fracture aptitude means a minimum of the material strength: a sum of constant load and frequency related load must lead to fracture.

We should remember that the Planck theory introduces the resonance wave frequency related to the black body radiation effect; basing on this approach we might try to consider a material strength minimum, as related to such resonance frequency, under the static vibration parts of load. The parameter H introduced for the processes in a thermal equilibrium, can depend also on the other experimental conditions, like:

$$H = \left(S^{Strength} - S^{Static\ Load}\right)\bar{H} \tag{10.6}$$

where the reduced value, $\bar{H}$, depends on the static applied load (e.g., applied pressure); for $S^{Strength} - S^{Static\ Load} = 0$ we obtain $H = 0$.

More complicated relations for H shall be considered for a shear load with the applied additional shear oscillations of given frequency ν; such oscillations may have different nature: mechanical, electric or magnetic. We expect that under a constant load near the critical point, such additional perturbations with frequency ν_0 might lead to a fracture, while at frequencies greater or smaller any fracture will not appear.

Following the presented approach, we assume that the break energy under the given confining load and shears will depend on the applied additional frequency related load. Along a fracture, the breaks might propagate not with a phase velocity, $\upsilon = \nu\lambda = \frac{\varpi}{k}$; $k = 2\pi\frac{n}{\lambda}$, but according to a group velocity:

$$\upsilon^g = \frac{\partial\varpi}{\partial k} \tag{10.7}$$

and for momentum we obtain:

$$p = \frac{H}{\lambda}n \tag{10.8}$$

For the total energy we may write the relativistic formulae:

$$\begin{aligned} E^2 = m^2c^4 = m_0^2c^4 + p^4c^2 = m_0^2c^4 + m^2\upsilon^2c^2 \\ m^2c^2 = m_0^2c^2 + m^2\upsilon^2 \end{aligned} \tag{10.9}$$

and for our case we may write per analogy:

$$\begin{aligned} H^2\nu^2 &= (\Delta\rho)^2\left(\frac{\omega}{k}\right)^4 + (\Delta\rho)^2\left(\frac{\partial\omega}{\partial k}\right)^2\left(\frac{\omega}{k}\right)^2 \\ H^2\nu^2 &= (\Delta\rho)^2\left(\frac{\omega}{k}\right)^2\left\{\left(\frac{\omega}{k}\right)^2 + \left(\frac{\partial\omega}{\partial k}\right)^2\right\} \end{aligned} \tag{10.10}$$

We should remember here about some papers related to a static load and other periodic uploadings including mechanical or, e.g., electric and magnetic oscillations (see Chelidze et al. 2010).

We may also consider an advanced approach: quantum-like theory of fracture probability wave.

In the continual and coherent states, the strain waves propagate according to the Asymmetric Continuum Theory and release-rebound principle. We know the laws for the wave propagation for the strains and rotation strains, $\hat{E}_{ns}$ and $\breve{E}_{ns}$ (see Chap. 2).

A release-rebound process occurs due to any elementary fracture and leads to a monochromatic wave emission. However, any fracture process is not governed by the classic, and even asymmetric mechanics; we need to search for new rules to describe such processes.

However, when considering the fracture problems, many similarities to the quantum theory appear, also such that these searched laws should be related either to an elementary process or to a multitude of events (n).

Our final aim is to deduce the rules for fracture processes following the probabilistic quantum approach.

To this end, we apply again the important assumption that a wave emitted relates to the most probable fracture frequency, as shown in Fig. 10.1. Thus, we have proposed an expression for the emitted energy in an elementary fracture as given by the Planck-like law (cf., Eqs. 10.3 and 10.4) with the constant H having a value specific to the given material and its state, and ν_0 defined earlier as fracture frequency.

The fracture event occurs in-between the continual and cohernt states und requires a further analysis; again, we will follow some principles of quantum theory.

We assume a plane fracture system: on any plane perpendicular to the position vector, $\bar{r}$, we may have some fracture plane.

At the constant axial strains, $\hat{E}_{11}$, $\hat{E}_{22}$, $\hat{E}_{33}$, we define a shear vector, $\hat{E}_s = \left\{\hat{E}_{(23)}, \hat{E}_{(31)}, \hat{E}_{(12)}\right\}$ (at a constant load, we may achieve such representation using a special, off-diagonal coordinate system; an invariant definition for such a vector is possible when using the 4D approach (cf., Teisseyre 2011) and for the rotation strain vector we can always put: $\breve{E}_s = \left\{\breve{E}_{[23]}, \breve{E}_{[31]}, \breve{E}_{[12]}\right\}$.

Further, we will use for an individually emitted shear wave the Bessel solution $\hat{E}_s(r, \varphi, z)$, related to a local coordinate system on an individual fracture plane perpendicular to the position vector, $\bar{z}$; we may define $\hat{E}_r^{\bar{z}}$, $\hat{E}_\varphi^{\bar{z}}$ as:

$$\hat{E}_s(r, \varphi, z) \to \left\{\hat{E}_r^{\bar{z}}(r, \varphi), \hat{E}_\varphi^{\bar{z}}(r, \varphi)\right\} \tag{10.11a}$$

and similarly for rotation

$$\breve{E}_s(r, \varphi, z) \to \left\{\breve{E}_r^{\bar{z}}(r, \varphi), \breve{E}_\varphi^{\bar{z}}(r, \varphi)\right\} \tag{10.11b}$$

Of course, we might consider a multitude of related fractures.

The elementary fracture process combining shear and rotation releases should relate the local coordinate systems with the z-axis oriented parallel to these vectors, $\hat{E}_s$ and $\breve{E}_s$.

Now, with these assumptions we may postulate the transitions from the generalized Sommerfeld conditions to those modified for the fracture process on any of the considered fracture planes; we write the generalized expressions related to the shear and rotation strain fields (Teisseyre 2009, 2011):

- for the shears:

$$\oint p_s \mathrm{d}q_s = n_s h \rightarrow \left\{ m \oint \frac{\partial \hat{E}_r^z}{\partial t} \sin(2\varphi) r \mathrm{d}\varphi,\ m \oint \frac{\partial \hat{E}_\varphi^z}{\partial t} \cos(2\varphi) r \mathrm{d}\varphi \right\} = \{n_r H,\ n_\varphi H\} \tag{10.12a}$$

- for the rotations:

$$\oint p_s \mathrm{d}q_s = n_s h \rightarrow \left\{ m \oint \frac{\partial \bar{E}_r^z}{\partial t} \sin(2\varphi) r \mathrm{d}\varphi,\ m \oint \frac{\partial \bar{E}_\varphi^z}{\partial t} \cos(2\varphi) r \mathrm{d}\varphi \right\} = \{n_r H,\ n_\varphi H\} \tag{10.12b}$$

where we have assumed the same numbers $\{n_r,\ n_\varphi\}$ for the correlated shear and rotation motions.

Following the quantum theory, these relations for the fracture processes, and that for fracture energy release (cf., Eq. 10.4), permit to apply the Schrödinger-like relation for the wave probability function, Ψ, related, here, to fracture processes in a given material and with a potential V(r,t):

$$\frac{iH}{2\pi} \frac{\partial \Psi(r,\ t)}{\partial t} = \left[-\frac{H^2}{8\pi^2 m} \Delta + V(r,\ t) \right] \Psi(r,\ t) \tag{10.13a}$$

However, the wave function, $\Psi(r,\ t)$, is not an invariant 3D function and therefore should be multiplied by the phase factor; when considering a system moving with a velocity $\upsilon(r' = r - \upsilon t,\ t' = t)$:

$$\Psi'(r',\ t') = \exp\left(-im \sum \upsilon_s r/\hbar + im\, \upsilon^2 t/2\hbar^2 \right) \Psi(r,t) \tag{10.13b}$$

we arrive again at the Schrödinger-like equation.

Moreover, we arrive at the transformation of energy and momentum:

$$E' = E + m\upsilon^2/2 - \sum p_s \upsilon_s, \quad p'_k = p_k - m\upsilon_k \tag{10.13c}$$

The potential function, $V(r,\ t)$, should be defined by the local conditions for strains and other fields; we may propose it as given by the local shear and rotation energies, Ê, Ē, for the given constant confining load, $\mathrm{E} = \frac{1}{3} \sum\limits_s E_{ss}$:

$$V(r,\ t) = 2\mu \left(\hat{\mathrm{E}} + \bar{\mathrm{E}} \right) \tag{10.14}$$

We assume that an individual elementary fracture process joins the shear fracture release with the rebound-release of rotation according to the Asymmetric Continuum Theory; it means that fracture probability function, $\Psi(r, t)$, should relate to a combined shear-rotation fracture event or to a complex of such events.

In our approach to fracture processes we assume that the Schrödinger-like equation describes the position probability of the a point-fracture event inside a continuum (instead of a point in the quantum theory).

Double fracture processes. A given body may have a double (or more) fracture strength values related to the double grain structure.

Instead of the Planck structures with the different strength values at a given constant pressure (Eq. 10.4), we write similar expressions for the radiated energy in a thermal equilibrium inside a fracture domain per unit wavelength and for the double strength levels, H_1, H_2:

$$e_1(\lambda) = \frac{8\pi v H_1}{\lambda^4}\left(\exp\left(\frac{vH_1}{KT}\right) - 1\right)^{-1}, \quad e_2(\lambda) = \frac{8\pi v H_2}{\lambda^4}\left(\exp\left(\frac{vH_2}{KT}\right) - 1\right) \tag{10.15a}$$

For the total energy, $e(\lambda) = e_1(\lambda) + e_2(\lambda)$, we can write:

$$e(\lambda) = \frac{8\pi v}{\lambda^4}\left(\frac{H_1}{\exp(vH_1/KT) - 1} + \frac{H_2}{\exp(vH_2/KT) - 1}\right) \tag{10.15b}$$

where λ is the radiated wavelength, K is the Boltzmann constant, T is the temperature, V is the elastic wave velocity, and characteristic material constants, H_1 and H_2, are related to the material-own binding energies.

The examples of the related radiated energies, $e(\lambda) = e_1(\lambda) + e_2(\lambda)$, might be shown on a figure similar to Fig. 10.1 for different values of H_1 and H_2.

Following the Schrödinger-like relations (10.13) and (10.14) we may write the probability functions of the fracture processes separately for the different grains, Ψ_1 and Ψ_2:

$$\frac{iH_1}{2\pi}\frac{\partial\Psi_1(r, t)}{\partial t} = \left[-\frac{H_1^2}{8\pi^2 m_1}\Delta + V(r, t)\right]\Psi_1(r, t), \quad V = 2\mu\left(\hat{\mathrm{E}} + \breve{\mathrm{E}}\right) \tag{10.16a}$$

$$\frac{iH_2}{2\pi}\frac{\partial\Psi_2(r, t)}{\partial t} = \left[-\frac{H_2^2}{8\pi^2 m_2}\Delta + V(r, t)\right]\Psi_2(r, t), \quad V = 2\mu\left(\hat{\mathrm{E}} + \breve{\mathrm{E}}\right) \tag{10.16b}$$

In relation to the electro-magnetic phenomena we might consider the Klein–Gordon-like relativistic relations:

$$\frac{1}{c^2}\left(\frac{\partial}{\partial t} + \frac{\mathrm{i}e}{\hbar}\Phi\right)^2\Psi(r, t) - \left(\sum_S \frac{\partial}{\partial x_s} + \frac{\mathrm{i}e}{\hbar}\mathrm{A}_s\right)^2\Psi(r, t) + \left(\frac{mc}{\hbar}\right)^2\Psi(r, t) = 0 \tag{10.17}$$

where Φ, A_s are the electro-magnetic potentials.

In particular, we might consider here the Schottky and Frenkel defects, already mentioned at the end of Chap. 8, in relation to the fracture processes (cf., Chaps. 5 and 9).

Without electro-magnetic interaction we obtain the pure Klein–Gordon-like equation as the relativistic Schrödinger-like equation.

We may conclude that, when considering and analyzing models of fracture processes in seismology, there may appear a question of the resonance frequency in destruction of materials.

Our considerations have been based on some important similarities between the backgrounds of the quantum theory (Q) and some principles suitable to build a basement for the fracture processes (F), especially those related to seismic events. We could list the main similarities for the quantum, Q, and fracture, F, analogies:

- (Q1) Appearance of the top frequency in radiation under a thermodynamic equilibrium (the black body radiation—a first fundamental problem leading to quantum mechanics);
- (F1) Assumption that the fracture processes run in the thermodynamic equilibrium within the related fracture recess and at the material strength minimum, as related to the resonance frequency under the constant pressure and additional shear vibration load; such a resonance frequency should be specific for a given material and its state related also to a given pressure;
- (Q2) Radiated energy per unit wavelength (monochromatic wave emission) is described by the Planck relation with the Planck constant, h;
- (F2) Assumption that there exists a similar form for a monochromatic wave emission with constant H specific for a given material and its related state, moreover, that a monochromatic wave emission is related to the release-rebound principle joining the shear and radiation fields in an elementary fracture process;
- (Q3) The quantum theory laws could be related either to a single process or to a multitude of events (n);
- (F3) Assumption that the proposed fracture theory could be based on an analogy that searched fracture laws should be related either to an elementary process or to a multitude of fracture events (n);
- (Q4) The theory is based on the generalized Sommerfeld conditions for momentum and moment of momentum, and
- (F4) Assumption that a fracture theory applies to the correlated fields as the time derivatives of shear and rotation strains; however, we additionally assume that these two fields remain mutually correlated by the release-rebound principle.

We have concluded that the presented similarities give grounds for formulating the Schrödinger-like relation for the wave probability function, Ψ, related, here, to fracture processes in a given material and pressure conditions.

We can state, finally, as follows:

- (Q5) The quantum theory is based on the the Schrödinger equation, and
- (F5) Assumption that the fracture theory could be based on the similar relations for the fracture probability function combing the shear-rotation fracture events

or a multitude of such events; the resulting probability conditions should depend on the observed intensities of the combined shear and rotation fields.

The last remark gives grounds for an urgent need of the word-wide system of the strain and rotation sensors with frequency characteristics reaching the long waves.

References

Chelidze T, Matcharashvili T, Lursmanashvili O, Varamashvili N, Zhukova E, Meparidze E (2010) Triggering and synchronization of stick-sip: experiments on spring-slider system. In: De Rubeis V, Czechowski Z, Teisseyre R (eds) Synchronization and triggering: from fracture to earthquake processes. Springer-Verlag, Berlin Heidelberg. p 123–164

Teisseyre R (2009) Tutorial on new development in physics of rotation motions. Bull Seismol Soc Am 99(2B):1028–1039

Teisseyre R (2011) Why rotation seismology: confrontation between classic and asymmetric theories. Bull Seismol Soc Am 101(4):1683–1691

Teisseyre R, Górski M (2012) Induced strains and defect continuum theory: internal reorganization of load. Acta Geophys 60(1):24–42

Chapter 11
Extreme Processes

11.1 Introduction

First, we consider an extension of our theory for the asymmetric fluids with the time rates of strains including rotation strains; such a theory can explain different extreme phenomena related to fluid dynamics, among others those usually known as the soliton waves. At an interface between liquid and gas we consider the propagation of surface waves and vortex motions. On a surface of liquid drops the rotation and shear strains become real, not molecular. Among other phenomena we consider a relation between the stress rates and the formation of dynamic viscous-dislocations valid for an incompressible fluid; the 2D turbulence structure leads us even to some molecular structures. Considering a solid continuum we return to fracture processes; the fracture phenomena progress with the simultaneous changes of material properties, including the granulation processes and material crushing; the constitutive laws undergo considerable changes leading to a narrow mylonite zone adjacent to fracture, where the time-rates of strains appear important. We arrive at some synchronization between a progressive fracture process and wave emission. Thus, in this chapter we present the asymmetric continuum theory including different types of material states: from the fluid state up to elastic continuum and the granulated/crushed material.

11.2 Extreme Processes

Fluids: Vortices and Phase Transitions. A mixture formed by the mutual relations of liquid and gas densities may be used to describe different processes when we additionally include a transition between these fluid states. We may include mutual transitions between these states by some operators very sensitive to temperature and gravity acceleration. With these aims we can describe the linear turbulence and vortex turbulence as extreme cases (Beirão da Veiga 2007; Teisseyre 2008a, b; Teisseyre and Górski 2009a, b; Teisseyre and Górski 2009b).

R. Teisseyre and M. Teisseyre-Jeleńska, *Asymmetric Continuum*,
GeoPlanet: Earth and Planetary Sciences, DOI: 10.1007/978-3-642-31860-3_11,

A special case presents a liquid surface zone and a gas above it; the molecular forces related to the surface tension decrease with temperature. The liquid (water) drops in a gas region play an important role. A surface tension depends also strongly on a surface curvature and thus we may expect a formation of liquid drops.

At an interface between liquid and gas we may consider the propagation of surface waves; here also an important role is played by surface tension. We may consider propagation of these waves in the X-direction along the surface plane near Z = 0. We can also introduce another coordinate system bound to the wave surface, ξ, η, ς at $\eta \equiv$ Y, and then we may write for the surface waves:

$$\chi = X + \alpha\cos(X - \varpi t), \quad \eta = Y, \quad \varsigma = Z + \gamma\sin(X - \varpi t), \tag{11.1a}$$

The reference coordinates (X, Y, Z) help us to define the variable curvilinear plane related to the surface waves and dividing gas and liquid states:

$$\varsigma = 0 \to Z + \gamma\sin(X - \varpi t) = 0 \tag{11.1b}$$

In Chap. 3 we have discussed the combined float transports; here, we continue those considerations on interaction between the different transport and float fields in more extreme conditions and including also the variations of density. Following the considered equations, e.g., Eq. 3.1, we may write for the transport and float fields, interconnected here, $\bar{\upsilon}_k$ and $\tilde{\upsilon}_k$:

$$\frac{\mathrm{d}\rho(\tilde{\upsilon}_k + \bar{\upsilon}_k)}{\mathrm{d}t} = \frac{\partial\rho(\tilde{\upsilon}_k + \bar{\upsilon}_k)}{\partial t} + \sum_s (\tilde{\upsilon}_s + \bar{\upsilon}_s)\frac{\partial\rho(\tilde{\upsilon}_k + \bar{\upsilon}_k)}{\partial x_s} = \eta\Delta(\tilde{\upsilon}_k + \bar{\upsilon}_s) + \tilde{F}_{ks} + \bar{F}_s$$

at $\bar{\upsilon}_s$ = constant and $\bar{F}_s \approx 0$:

$$(\tilde{\upsilon}_k + \bar{\upsilon}_k)\frac{\partial\rho}{\partial t} + \rho\frac{\partial\tilde{\upsilon}_k}{\partial t}\tilde{\upsilon}_k + \rho\sum_s (\tilde{\upsilon}_s + \bar{\upsilon}_s)\left[\frac{\partial\tilde{\upsilon}_k}{\partial x_s} + (\tilde{\upsilon}_k + \bar{\upsilon}_k)\frac{\partial\rho}{\rho\partial x_s}\right] = \eta\Delta\tilde{\upsilon}_k + \tilde{F}_{ks} \tag{11.2a}$$

We may consider a complex case for a gas medium with some counterpart of the fluid (fluid drops formed due to a surface tension); inside this transition zone we may redefine the continuity relation for the gas density, $\tilde{\rho}$; we may assume for the liquid part: $\bar{\rho}$ = const. Then, we may write the continuity relation:

$$\begin{aligned} &\frac{d(\bar{\rho} + \tilde{\rho})}{dt} = \frac{\partial(\bar{\rho} + \tilde{\rho})}{\partial t} + \sum_s \frac{\partial\bar{\rho}(\tilde{\upsilon}_s + \bar{\upsilon}_s)}{\partial x_s} + \sum_s \frac{\partial\tilde{\rho}(\tilde{\upsilon}_s + \bar{\upsilon}_s)}{\partial x_s} = 0 \\ &\frac{\partial\tilde{\rho}}{\partial t} + (\bar{\rho} + \tilde{\rho})\sum_s \frac{\partial\tilde{\upsilon}_s}{\partial x_s} + \sum_s (\tilde{\upsilon}_s + \bar{\upsilon}_s)\frac{\partial\tilde{\rho}}{\partial x_s} \approx 0 \end{aligned} \tag{11.2b}$$

Instead of (8.2a) we get, under these conditions, a relation for the gas region:

$$\frac{\mathrm{d}\tilde{\rho}(\tilde{v}_k + \bar{v}_k)}{\mathrm{d}t} = \frac{\partial\tilde{\rho}(\tilde{v}_k + \bar{v}_k)}{\partial t} + \sum_s (\tilde{v}_s + \bar{v}_s)\frac{\partial\tilde{\rho}(\tilde{v}_k + \bar{v}_k)}{\partial x_s} = \tilde{\eta}\Delta\tilde{v}_k + \tilde{F}_k \rightarrow$$

$$(\tilde{v}_k + \bar{v}_k)\frac{\partial\tilde{\rho}}{\partial t} + \tilde{\rho}\frac{\partial\tilde{v}_k}{\partial t}\tilde{v}_k + \tilde{\rho}\sum_s (\tilde{v}_s + \bar{v}_s)\left[\frac{\partial\tilde{v}_k}{\partial x_s} + (\tilde{v}_k + \bar{v}_k)\frac{\partial\tilde{\rho}}{\tilde{\rho}\partial x_s}\right] = \eta\Delta\tilde{v}_k + \tilde{F}_{ks} \tag{11.3a}$$

or with (11.2b):

$$\begin{gathered}\frac{\partial\tilde{\rho}}{\partial t} + (\bar{\rho} + \tilde{\rho})\sum_s \frac{\partial\tilde{v}_s}{\partial x_s} + \sum_s (\tilde{v}_s + \bar{v}_s)\frac{\partial\tilde{\rho}}{\partial x_s} \approx 0 \\ \frac{\partial(\tilde{\rho}\tilde{v}_k)}{\partial t} + \sum_s (\tilde{v}_s + \bar{v}_s)\frac{\partial(\tilde{\rho}\tilde{v}_k)}{\partial x_s} = \tilde{\eta}\Delta\tilde{v}_k + \tilde{F}_k\end{gathered} \tag{11.3b}$$

and in this way we obtain a joint relation:

$$\tilde{\rho}\frac{\partial\tilde{v}_k}{\partial t} - \tilde{v}_k\tilde{\rho}\sum_s \frac{\partial\tilde{v}_s}{\partial x_s} + \tilde{\rho}\sum_s (\tilde{v}_s + \bar{v}_s)\frac{\partial\tilde{v}_k}{\partial x_s} + \tilde{v}_k\sum_s \bar{v}_s\frac{\partial\tilde{\rho}}{\partial x_s} = \tilde{\eta}\Delta\tilde{v}_k + \tilde{F}_k \tag{11.4}$$

We should also note that the time derivative of pressure relates to the time derivative of density; thus, for an ideal and adiabatic gas medium, we write for the acoustic pressure:

$$\frac{\partial\tilde{p}}{\partial t} = \tilde{c}^2\frac{\partial\tilde{\rho}}{\partial t};\ \frac{\partial\tilde{p}}{\partial\tilde{\rho}} = \tilde{c}^2 \text{ and } \tilde{c}^2 = \frac{d\tilde{p}}{d\tilde{\rho}} = \frac{C_P\tilde{p}}{C_V\tilde{\rho}} \tag{11.5a}$$

where for the difference between a total and a variable part of pressure, $\tilde{p}$, and a mean pressure p (the variable part is called acoustic pressure), we can write similarly to Eq. 2.4a (Chap. 2) the equation for the acoustic pressure:

$$\sum_s \frac{\partial^2\tilde{p}}{\partial x_s\partial x_s} - \frac{1}{\tilde{c}^2}\frac{\partial^2\tilde{p}}{\partial t^2} = 0; \tag{11.5b}$$

where:

$$\tilde{p} - p = \frac{\partial\tilde{p}}{\partial\tilde{\rho}}(\tilde{\rho} - \rho), \quad \tilde{c} = \sqrt{\frac{B}{\rho}}, \quad B = \rho\left(\frac{\partial\tilde{p}}{\partial\tilde{\rho}}\right)_{\text{adiabatic}}, \quad \tilde{p} = B\frac{(\tilde{\rho} - \rho)}{\rho}$$

and the compressibility $\beta = \frac{1}{B} = -\frac{1}{V}\frac{\partial V}{\partial p}$.

Finally we arrive at

$$\frac{\partial\tilde{p}}{\partial t} = \tilde{c}^2\frac{\partial\tilde{\rho}}{\partial t};\ \frac{\partial\tilde{p}}{\partial\tilde{\rho}} = \tilde{c}^2 \text{ and } \tilde{c}^2 = \frac{d\tilde{p}}{d\tilde{\rho}} = \frac{C_P\tilde{p}}{C_V\tilde{\rho}} \tag{11.5c}$$

where $\tilde{c}$ is the sound velocity and C_p and C_v are the specific heat at constant pressure and volume.

Further on, the sound waves, $\frac{\partial\tilde{p}}{\partial t}$, can be neglected in our considerations.

Using the vortex motion Eqs. 4.23 and 4.24 (see Chap. 4), for a stationary vortex we have the following equations for the fields $\tilde{r}$ and $\tilde{z}$ at $\tilde{\omega} = \text{const.}$:

$$\begin{aligned}\frac{\partial(\rho\tilde{r}^2)}{\partial r} + \frac{\partial(\rho\tilde{\omega}\tilde{r})}{\partial\varphi} + \frac{\partial(\rho\tilde{r}\tilde{z})}{\partial z} &= \tilde{F} \\ \frac{\partial(\rho\tilde{r}\tilde{z})}{\partial r} + \frac{\partial(\rho\tilde{\omega}\tilde{z})}{\partial\varphi} + \frac{\partial(\rho\tilde{z}^2)}{\partial z} &= \tilde{N}\end{aligned} \tag{11.6a}$$

where $x_s \to (r, \varphi, z); \upsilon_s \to (\tilde{r}, \tilde{\varphi} \approx 0, \tilde{z})$.

We obtain the following solution:

$$\tilde{r} = \tilde{r}_0 \exp(\alpha r - \alpha z), \qquad \tilde{z} = \tilde{z}_0 \exp(\alpha r - \alpha z) \tag{11.6b}$$

For the considered solution for a vortex motion (see Chap. 4, Eq. 4.33) with the boundary conditions $z = 0$ and $R = R_0$ ($r = 0$ for $R = R_0 - r$), we obtain the solution:

$$\begin{aligned}&\tilde{r} \approx \tilde{r}_0 \exp(\alpha(R_0 - R) - \alpha z), \tilde{\omega} \approx \tilde{\omega}_0 \exp(\alpha(R - R_0) - \alpha z), \\ &\tilde{z} \approx \tilde{z}_0 \exp(\alpha(R - R_0) - \alpha z)\end{aligned} \tag{11.6c}$$

A vortex is thus described as follows (as visualized in Fig. 11.1):

$$\begin{aligned}&z = 0 : R = R_0, (r = 0), \tilde{r} = \tilde{r}_0 \exp[\alpha(R_0 - R - z)] \to \tilde{r} = \tilde{r}_0 \\ &z = Z_0 : R = 0, (r = R_0), \tilde{r} = \tilde{r}_0 \exp[\alpha(R_0 - R - z)] \to \tilde{r} = \tilde{r}_0 \exp[\alpha(R_0 - Z_0)]\end{aligned} \tag{11.6d}$$

The vortex transport motions can induce the molecular strains (cf., Chap. 4, Eqs. 4.8a, b);

$$\begin{aligned}&\tilde{E}_{kl} = \tilde{E}_{(kl)} + \tilde{E}_{[kl]}, \quad \tilde{S}_{(kl)} = \tilde{\eta}\tilde{E}_{(kl)}, \quad \tilde{S}_{[kl]} = \tilde{\eta}\tilde{E}_{[kl]} \\ \text{and} \quad &\frac{1}{3}\sum_s \tilde{S}_{ss} = -\frac{\partial p}{\partial t} = \frac{1}{3}k\sum_s \tilde{E}_{ss} \approx 0\end{aligned} \tag{11.7a}$$

where according to (11.6c):

$$\begin{aligned}\tilde{E}_{(kl)} &= \frac{1}{2}\left(\frac{\partial \upsilon_l}{\partial x_k} + \frac{\partial \upsilon_k}{\partial x_l}\right), \tilde{E}_{[kl]} = \frac{1}{2}\left[\frac{\partial \upsilon_l}{\partial x_k} - \frac{\partial \upsilon_k}{l}\right] ; x_s \to (R, \varphi, z); \upsilon_s \to (\tilde{r}, \tilde{\varphi} \approx \text{const}, \tilde{z}) \\ \tilde{E}_{(zR)} &= \frac{1}{2}\left(\frac{\partial \tilde{r}}{\partial z} + \frac{\partial \tilde{z}}{\partial R}\right) \approx \left(\frac{\tilde{z}_0}{2} - \frac{\tilde{r}_0}{2}\right)\alpha \exp(-\alpha R_0)\exp(\alpha(R - z)), \\ \tilde{E}_{[zR]} &= \frac{1}{2}\left[\frac{\partial \tilde{r}}{\partial z} - \frac{\partial \tilde{z}}{\partial R}\right] \approx -\left(\frac{\tilde{r}_0}{2} + \frac{\tilde{z}_0}{2}\right)\alpha \exp(-_0)\exp(\alpha(R - z))\end{aligned} \tag{11.7b}$$

The obtained expressions for the molecular strains, $\tilde{E}_{(zR)}$ and $\tilde{E}_{[zR]}$, bring the possible strain values for $z = 0$ and $z = Z_0$:

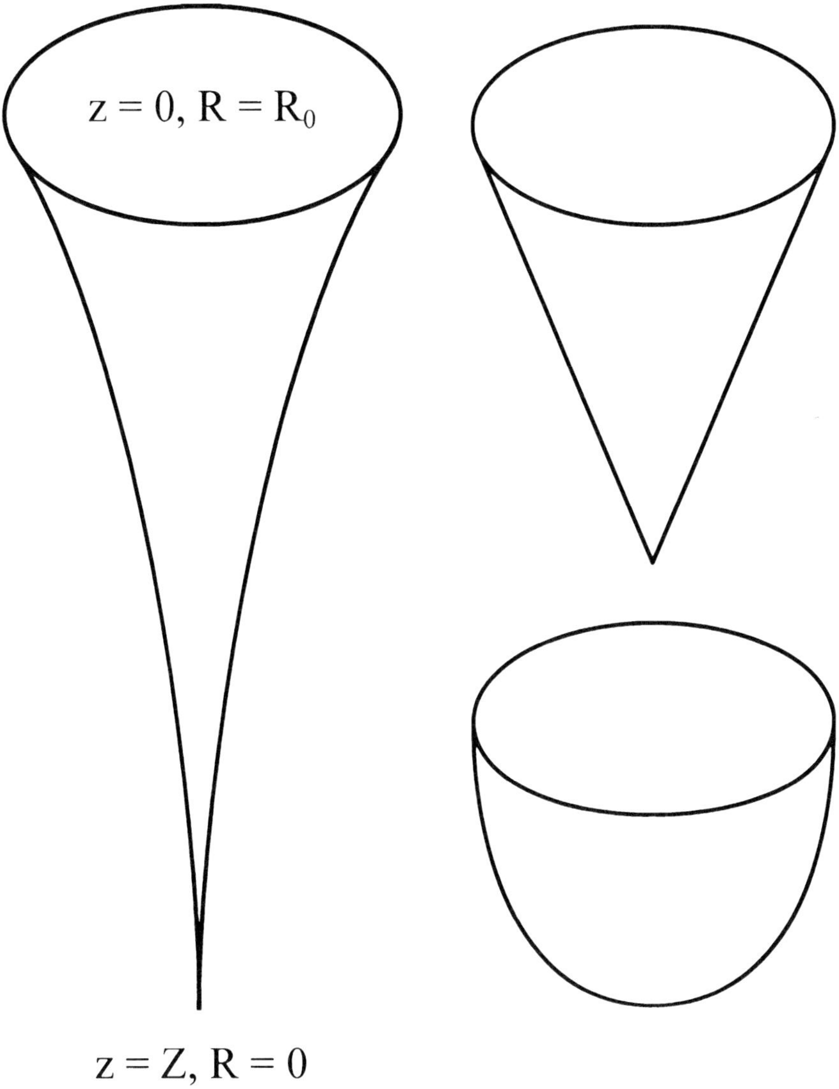

Fig. 11.1 The stationary vortex flows:, *left figure* accelerating downward for $R_0 - Z_0 > 0$ and increasing as $\tilde{r} = \tilde{r}_0 \exp[\alpha(R_0 - Z_0)]$ ↑, *right upper one* laminar with constant velocity $R_0 - Z_0 = 0$ and increasing as $\tilde{r} = \tilde{r}_0$, *right lower one* viscous slowing down flow for $R_0 - Z_0 < 0$ and $\tilde{r} = \tilde{r}_0 \exp[\alpha(R_0 - Z_0)]$ ↓

$$\begin{aligned}
\tilde{E}_{(z=0,R=0)} &\approx \left(\frac{\tilde{z}_0}{2} - \frac{\tilde{r}_0}{2}\right)\alpha\exp(-\alpha R_0), \quad E_{[z=0,R=0]} \approx -\left(\frac{\tilde{r}_0}{2} + \frac{\tilde{z}_0}{2}\right)\alpha\exp(-\alpha R_0) \\
\tilde{E}_{(z=Z_0,R=R_0)} &\approx \left(\frac{\tilde{z}_0}{2} - \frac{\tilde{r}_0}{2}\right)\alpha\exp(-\alpha Z_0), \quad E_{[z=Z_0,R=R_0]} \approx -\left(\frac{\tilde{r}_0}{2} + \frac{\tilde{z}_0}{2}\right)\alpha\exp(-\alpha Z_0)
\end{aligned} \tag{11.8a}$$

The described features of the vortex motions explain many observed vortex forms.

Now we propose to repeat the above-mentioned consideration with the time dependent fields. According to Eqs. 4.23 and 4.24 (see. Chap. 4) for a stationary vortex we have the equations for the fields, $\tilde{r}$ and $\tilde{z}$ at $\tilde{\omega} = \text{const.}$, where $x_s \to (r, \varphi, z); \upsilon_s \to (\tilde{r}, \tilde{\varphi} \approx 0, \tilde{z})$,

$$\begin{aligned} \frac{\partial(\rho\tilde{r}^2)}{\partial r} + \frac{\partial(\rho\tilde{\omega}\tilde{r})}{\partial\varphi} + \frac{\partial(\rho\tilde{r}\tilde{z})}{\partial z} = \tilde{F} \\ \frac{\partial(\rho\tilde{r}\tilde{z})}{\partial r} + \frac{\partial(\rho\tilde{\omega}\tilde{z})}{\partial\varphi} + \frac{\partial(\rho\tilde{z}^2)}{\partial z} = \tilde{N} \end{aligned} \quad (11.9a)$$

leading to the solution:

$$\tilde{r} = \tilde{r}_0 \exp(\alpha r - \alpha z), \; \tilde{z} = \tilde{z}_0 \exp(\alpha r - \alpha z) \quad (11.9b)$$

For a vortex $R = R_0 - r$ we write $\gamma(R_0 - R) = \gamma z$ (we have a boundary condition at $r = 0$) and we obtain:

$$\tilde{r} \approx \tilde{r}_0 \exp(-\alpha R_0) \exp(\alpha(R - z)), \quad \tilde{z} \approx \tilde{z}_0 \exp(-\alpha R_0) \exp(\alpha(R - z)) \quad (11.9c)$$

A vortex at $R = R_0 - r$ is thus described as follows (Fig. 11.1):

$$\begin{aligned} &z = 0 : R = R_0, (r = 0), \tilde{r} = \tilde{r}_0 \exp[\alpha(R_0 - R - z)] \to \tilde{r} = \tilde{r}_0 \\ &z = Z_0 : R = 0, (r = R_0), \tilde{r} = \tilde{r}_0 \exp[\alpha(R_0 - R - z)] \to \tilde{r} = \tilde{r}_0 \exp[\alpha(R_0 - Z_0)] \end{aligned} \quad (11.9d)$$

The vortex transport motions can induce molecular strains, Eq. 11.7 (cf., Chap. 4, Eqs. 4.8a, b):

$$\begin{aligned} &\tilde{E}_{kl} = \tilde{E}_{(kl)} + \tilde{E}_{[kl]}, \; \tilde{S}_{(kl)} = \tilde{\eta}\tilde{E}_{(kl)}, \; \tilde{S}_{[kl]} = \tilde{\eta}\tilde{E}_{[kl]} \\ &\text{and } \frac{1}{3}\sum_s \tilde{S}_{ss} = -\frac{\partial p}{\partial t} = \frac{1}{3}k\sum_s \tilde{E}_{ss} \approx 0 \end{aligned} \quad (11.10a)$$

where according to (11.6c):

$$\begin{aligned} \tilde{E}_{(kl)} &= \frac{1}{2}\left(\frac{\partial\upsilon_l}{\partial x_k} + \frac{\partial\upsilon_k}{\partial x_l}\right), \; \tilde{E}_{[kl]} = \frac{1}{2}\left[\frac{\partial\upsilon_l}{\partial x_k} - \frac{\partial\upsilon_k}{\partial x_l}\right]; \; x_s \to (R, \varphi, z); \upsilon_s \to (\tilde{r}, \tilde{\varphi} \approx \text{const}, \tilde{z}) \\ \tilde{E}_{(zR)} &= \frac{1}{2}\left(\frac{\partial\tilde{r}}{\partial z} + \frac{\partial\tilde{z}}{\partial R}\right) \approx \left(\frac{\tilde{z}_0}{2} - \frac{\tilde{r}_0}{2}\right)\alpha\exp(-\alpha R_0)\exp(\alpha(R - z)), \\ \tilde{E}_{[zR]} &= \frac{1}{2}\left[\frac{\partial\tilde{r}}{\partial z} - \frac{\partial\tilde{z}}{\partial R}\right] \approx -\left(\frac{\tilde{r}_0}{2} + \frac{\tilde{z}_0}{2}\right)\alpha\exp(-\alpha R_0)\exp(\alpha(R - z)) \end{aligned} \quad (11.10b)$$

The obtained expressions for the molecular strains, $\tilde{E}_{(zR)}$ and $\tilde{E}_{[zR]}$, bring the possible strain values for $z = 0$ and $z = Z_0$:

$$\tilde{E}_{(z=0,R=0)} \approx \left(\frac{\tilde{z}_0}{2} - \frac{\tilde{r}_0}{2}\right)\alpha\exp(-\alpha R_0), \quad \breve{E}_{[z=0,R=0]} \approx -\left(\frac{\tilde{r}_0}{2} + \frac{\tilde{z}_0}{2}\right)\alpha\exp(-\alpha R_0)$$
$$\tilde{E}_{(z=Z_0,R=R_0)} \approx \left(\frac{\tilde{z}_0}{2} - \frac{\tilde{r}_0}{2}\right)\alpha\exp(-\alpha Z_0), \quad \breve{E}_{[z=Z_0,R=R_0]} \approx -\left(\frac{\tilde{r}_0}{2} + \frac{\tilde{z}_0}{2}\right)\alpha\exp(-\alpha Z_0) \tag{11.10c}$$

The described features of the vortex motions explain many observed vortex forms.

Of course in some way the molecular strains are related to existing transport motions; the fluid point motions have six degrees of freedom: displacement velocities and rotations.

However, we should change the equations when taking into account that a gas medium with some volume part related to the liquid drops might be defined, at some strong simplification, as a mixed continuum with percentage of gas density, (%), and liquid density (1 −%); for such a mixed continuum we may use these percentages for the density and transport equations, e.g.:

$$\frac{\partial\tilde{\rho}}{\partial t}(\%) + \bar{\rho}[(1-\%) + \tilde{\rho}(\%)]\sum_s \frac{\partial\tilde{\upsilon}_s}{\partial x_s} + \sum_s(\tilde{\upsilon}_s + \bar{\upsilon}_s)\frac{\partial\tilde{\rho}}{\partial x_s}(\%) \approx 0 \qquad ((11.11a))$$

Similarly we write some other relations at a constant fluid density, $\bar{\rho} = \text{const.}$

$$\begin{aligned}&(\tilde{\upsilon}_k + \bar{\upsilon}_k)\frac{\partial\tilde{\rho}(\%)}{\partial t} + \tilde{\rho}(\%)\frac{\partial\tilde{\upsilon}_k}{\partial t} + \tilde{\upsilon}_k\sum_s(\tilde{\upsilon}_s + \bar{\upsilon}_s)\frac{\partial\tilde{\rho}(\%)}{\partial x_s} + \\ &[\bar{\rho}(1-\%) + \tilde{\rho}(\%)]\sum_s(\tilde{\upsilon}_s + \bar{\upsilon}_s)\frac{\partial\tilde{\upsilon}_k}{\partial x_s} = \tilde{\eta}\Delta\tilde{\upsilon}_k + \tilde{F}_k + \bar{F}_k\end{aligned} \tag{11.11b}$$

and when eliminating the $\frac{\partial\tilde{\rho}}{\partial t}(\%)$ term from (11.13a):

$$\begin{aligned}&\tilde{\rho}(\%)\frac{\partial\tilde{\upsilon}_k}{\partial t} - (\tilde{\upsilon}_k + \bar{\upsilon}_k)[\bar{\rho}(1-\%) + \tilde{\rho}(\%)]\sum_s\frac{\partial\tilde{\upsilon}_s}{\partial x_s} - (\tilde{\upsilon}_k + \bar{\upsilon}_k)\sum_s(\tilde{\upsilon}_s + \bar{\upsilon}_s)\frac{\partial\tilde{\rho}}{\partial x_s}(\%) + \\ &\tilde{\upsilon}_k\sum_s(\tilde{\upsilon}_s + \bar{\upsilon}_s)\frac{\partial\tilde{\rho}(\%)}{\partial x_s} + [\bar{\rho}(1-\%) + \tilde{\rho}(\%)]\sum_s(\tilde{\upsilon}_s + \bar{\upsilon}_s)\frac{\partial\tilde{\upsilon}_k}{\partial x_s} = \tilde{\eta}\Delta\tilde{\upsilon}_k + \tilde{F}_k + \bar{F}_k\end{aligned} \tag{11.11c}$$

or

$$\begin{aligned}&\tilde{\rho}(\%)\frac{\partial\tilde{\upsilon}_k}{\partial t} - (\tilde{\upsilon}_k + \bar{\upsilon}_k)[\bar{\rho}(1-\%) + \tilde{\rho}(\%)]\sum_s\frac{\partial\tilde{\upsilon}_s}{\partial x_s} - \bar{\upsilon}_k\sum_s(\tilde{\upsilon}_s + \bar{\upsilon}_s)\frac{\partial\tilde{\rho}(\%)}{\partial x_s} + \\ &+ [\bar{\rho}(1-\%) + \tilde{\rho}(\%)]\sum_s(\tilde{\upsilon}_s + \bar{\upsilon}_s)\frac{\partial\tilde{\upsilon}_k}{\partial x_s} = \tilde{\eta}\Delta\tilde{\upsilon}_k + \tilde{F}_k + \bar{F}_k\end{aligned} \tag{11.11d}$$

In a rough approximation we can solve separately the equations for gas only with a final result multiplied by "%", and separately assume that the solution for a fluid part "1−%" should lead to the same motions as for gas, but diminished by a ratio of densities, $\tilde{\rho}/\bar{\rho} << 1$, e.g., $\bar{\upsilon}_k = \tilde{\upsilon}_k \frac{\tilde{\rho}}{\bar{\rho}}$; moreover, to a fluid part we should add the vertical gravity effect, $\bar{\rho} g_z$. Here, we confine ourselves to remind only an example of such mixed floats, as has been presented in Chap. 5, Fig. 5.3.

We should take into account that on any liquid/gas surface, including surfaces of drops and bubbles, we expect the rotation tensions; the molecular strains—rotational and shear, become the real strains. We arrive to a more complicated thermodynamic conditions with gas evaporation and appearance of liquid drops.

However, further on we will not enter into the complicated flow relations (cf., Beirão da Veiga, 2007) and related thermodynamic conditions in a gas medium with its evaporation effects, cloud formation and appearance of drops.

At the end we mention a special case related to the two neighboring gas flows of opposite-direction. The oppositely oriented neighboring gas motions may lead us to a probability of the vortex formation and turbulence effects, especially in the regions at higher temperatures and thus low densities; this problem has already been considered in Chap. 4. However, here we may join such a probability with some quantum-like relations as discussed in Chap. 10; there we have discussed the fracture processes and fracture energy release with the help of the Schrödinger-like relation (cf., Eq. 10.4). Here, we will use similar probability relations for the formation of vortex motions. Thus, a wave probability function, Ψ, will relate to a vortex micro-motion appearance at some potential $V(r,t)$ and constant, $\tilde{H}$, depending among other factors on temperature and gas density:

$$\frac{i\tilde{H}}{2\pi}\frac{\partial\Psi(r,t)}{\partial t} = \left[-\frac{\tilde{H}^2}{8\pi^2 m}\Delta + V(r,t)\right]\Psi(r,t) \tag{11.12}$$

This relation may be used to describe a vortex path with $r \to r' - \upsilon t'$ and $t = t'$; when considering a system of the opposite-direction motions with a relative velocity υ, we may obtain a similar form of equations defining a related probability, but of course with another definition of the H constant, e.g., $H \to \tilde{H}$, describing an appearance of a vortex motion, instead of a fracture probability.

Returning to the flow relations (11.4), we might consider the vortex motions with its float due to a constant motion, $\bar{\upsilon}_s$.

For vortex motions at a constant float we can write the relations for a gas medium with variable density (see Eq. 3.2b):

$$\tilde{\rho}\frac{\partial\tilde{\upsilon}_k}{\partial t} - \tilde{\upsilon}_k\tilde{\rho}\sum_s \frac{\partial\tilde{\upsilon}_s}{s} + \tilde{\rho}\sum_s(\tilde{\upsilon}_s + \bar{\upsilon}_s)\frac{\partial\tilde{\upsilon}_k}{\partial x_s} + \tilde{\upsilon}_k\sum_s \bar{\upsilon}_s\frac{\partial\tilde{\rho}}{\partial x_s} = \tilde{\eta}\Delta\tilde{\upsilon}_k + \tilde{F}_k$$

$$\to \tilde{\rho}\frac{\partial\tilde{\upsilon}_k}{\partial t} + \tilde{\rho}\sum_s(\tilde{\upsilon}_s + \bar{\upsilon}_s)\frac{\partial\tilde{\upsilon}_r}{\partial x_s} + \tilde{\upsilon}_k\sum_s\left(\bar{\upsilon}_s\frac{\partial\tilde{\rho}}{\partial x_s} - \tilde{\rho}\frac{\partial\tilde{\upsilon}_s}{\partial x_s}\right) = \tilde{\eta}\Delta\upsilon_k + \tilde{F}_k \tag{11.13a}$$

and further on with $\partial\tilde{v}_\varphi/\partial\varphi \approx 0$:

$$\begin{aligned}
&\tilde{\rho}\frac{\partial\tilde{v}_r}{\partial t} + \tilde{\rho}\tilde{v}\frac{\partial\tilde{v}_r}{\partial r} - \tilde{\rho}\tilde{v}_r\left(\frac{\partial\tilde{v}_r}{\partial r} + \frac{\partial\tilde{v}_z}{\partial z}\right) = \tilde{\eta}\Delta\tilde{v}_r + \tilde{F}_r \\
&\tilde{\rho}\frac{\partial\tilde{v}_\varphi}{\partial t} + \tilde{\rho}\tilde{v}_r\frac{\partial\tilde{v}_{\varphi r}}{\partial r} + \tilde{\rho}\tilde{v}_z\frac{\partial\tilde{v}_{\varphi r}}{\partial z} - \tilde{\rho}\tilde{v}_\varphi\frac{\partial\tilde{v}_r}{\partial r} - \tilde{\rho}\tilde{v}_\varphi\frac{\partial\tilde{v}_z}{\partial z} = \tilde{\eta}\Delta\tilde{v}_\varphi + \tilde{F}_\varphi \\
&\tilde{\rho}\frac{\partial\tilde{v}_z}{\partial t} + \tilde{v}_s\frac{\partial\tilde{v}_z}{\partial z} + \tilde{v}_z - \tilde{\rho}\tilde{v}_z\frac{\partial\tilde{v}_r}{\partial r} - \tilde{\rho}\tilde{v}_z\frac{\partial\tilde{v}_z}{\partial z} = \tilde{\eta}\Delta\tilde{v}_z + \tilde{F}_z
\end{aligned} \tag{11.13b}$$

For a vortex liquid with a constant density we may write similar relations and for a liquid, e.g., water, boundary with gas, e.g., $F = g$.

Basing on these considerations we define the asymmetric fluids by the relations (Chap. 2):

$$\tilde{S}_{kl} = \tilde{S}_{(kl)} + \tilde{S}_{[kl]}, \quad \tilde{E}_{kl} = \tilde{E}_{(kl)} + \tilde{E}_{[kl]} \tag{11.14}$$

where the fields $\tilde{S}_{kl}$, $\tilde{E}_{kl}$, $\tilde{\omega}_{kl}$ mean the time rates of stresses, strains and rotations in fluids.

For an energy rate for fluids we write

$$\tilde{E} = \tilde{S}_{(kl)}\tilde{E}_{(kl)} + \tilde{S}_{[kl]}\tilde{E}_{[kl]} \tag{11.15}$$

while taking for the sake of simplicity an incompressible fluid, we put

$$\tilde{S}_{(kl)} = 2\eta^0\tilde{E}_{(kl)}\,, \quad \tilde{S}_{[kl]} = 2\eta^0\tilde{E}_{[kl]}\,, \tag{11.16}$$

where η^0 is a reference viscosity.

For fluids we may introduce the dynamical phase indexes, $\tilde{e}^0$ and $\tilde{\chi}^0$, and we put

$$\tilde{E}_{(ik)} = \tilde{e}^0\tilde{E}^0_{(ik)} = \tilde{e}^0\frac{1}{2}\left(\frac{\partial v_k}{\partial x_i} + \frac{\partial v_i}{\partial x_k}\right), \qquad \tilde{E}_{[ik]} = \tilde{\chi}^0\tilde{E}^0_{[ik]} = \tilde{\chi}^0\frac{1}{2}\left(\frac{\partial v_k}{\partial x_i} - \frac{\partial v_i}{\partial x_k}\right) \tag{11.17}$$

We shall note that the discussed wave equations and their specific solutions will remain fully valid for the fluids with the time rates of fields (everywhere in the derived relations we shall replace the discussed fields by their time rates). The introduced velocity field v shall obey the Navier–Stokes transport equation.

For the sake of simplicity we will consider, further on, the incompressible fluid ($\tilde{E}_{ss} = 0$, $\tilde{E}_{ik} = \tilde{E}^D_{ik}$) and put $\left|\tilde{e}^0\right| = 1$, $\tilde{\chi}^0 = 1$.

The extreme motion phenomena, related to the shear and spin rates, can be described, similarly to Eqs. 5.13 and 5.18 (Chap. 5), by the dynamic disclosure:

$$\begin{aligned}
B_l &= \oint E_{pl}dx_p = \varepsilon_{qns}\iint\frac{\partial E_{sl}}{\partial x_n}ds_q, \\
\tilde{B}_l &= \oint\left(\tilde{E}_{(kl)} + \tilde{\omega}_{[kl]}\right)\mathrm{d}l_k = \oint\left[\tilde{e}^0\tilde{E}^0_{(kl)} + \tilde{\omega}^0_{[kl]}\right]\mathrm{d}l_k,
\end{aligned} \tag{11.18}$$

and we can define the dynamical dislocations, v-dislocations:

$$\tilde{\alpha}_{pl} - \frac{1}{2}\tilde{\alpha}_{ss}\delta_{pl} = \varepsilon_{pmk}\frac{\partial}{\partial x_m}\left[\tilde{E}_{(kl)} + \tilde{\omega}_{[kl]}\right] \tag{11.19a}$$

$$\tilde{\alpha}_{pl} - \frac{1}{2}\tilde{\alpha}_{ss}\delta_{pl} = \frac{\varepsilon_{pmk}}{2\eta}\frac{\partial}{\partial x_m}\left((\tilde{S}_{(kl)} + \tilde{S}_{[kl]}\right) . \tag{11.19b}$$

We may also consider a special mechanism related to the solution for the molecular twist (shear strain rate) and spin (rotation strain rate); in a special coordinate system, we can present the strain rates by the common strain rate vectors, $\tilde{E}_{[s]}$ and $\tilde{E}_{(s)}$, mutually interrelated:

$$\tilde{E}_{[s]} = \mathrm{i}\hat{E}_{(s)}\ ; \quad \tilde{E}_{[s]} = \overset{0}{E}_{[s]} exp[\mathrm{i}(k_i x_i - \varpi t)]\ , \quad \hat{E}_{(s)} = \hat{E}^0_{(s)} exp[\mathrm{i}(k_i x_i - \varpi t)] \tag{11.20a}$$

For such a case we may define the complex solutions as follows:

$$\tilde{E}^0_{[s]} = \mathrm{i}\tilde{E}^0_{(s)} \tag{11.20b}$$

These six homogeneous conditions could be fulfilled also for the 2D solution in the systems $\{r,\ \varphi,\ z\}$:

$$\tilde{E}_{[r]} = \mathrm{i}\tilde{E}_{(r)}, \quad \tilde{E}_{[\varphi]} = \mathrm{i}\tilde{E}_{(\varphi)}, \quad \tilde{E}_{[z]} = \mathrm{i}\tilde{E}_{(z)} \tag{11.20c}$$

corresponding to a turbulence structure.

In some circumstances we may have a coincidence of the spin and twist rates; such a coincidence may lead to another type of the extreme motions: for the 2D plane wave solutions and propagation in the x-direction we obtain for a sum of these two motions with the amplitude A:

$$\tilde{E}_{(12)} + \tilde{E}_{[12]} \equiv \tilde{E}_{(12)} + \tilde{\omega}_{12} \rightarrow \frac{\partial \upsilon_2}{\partial x_1} \rightarrow \upsilon = exp[\mathrm{i}(kx - \varpi t)] \tag{11.21a}$$

and for the displacement velocity:

$$\upsilon \propto \Lambda exp[\mathrm{i}(kx - \varpi t)] . \tag{11.21b}$$

We find that under this condition of a full coincidence, the resulting wave amplitude becomes multiplied by a wavelength; for a long wave, Λ, this situation will lead to a large magnification of the respective wavelet amplitudes; e.g., we arrive at a soliton wave.

An extension of our theory for the asymmetric fluids has been based on the time rates of stresses, strains and rotations; such a theory can explain different extreme phenomena related to fluid dynamics, among others those usually known as the soliton waves.

Many other phenomena, in oceans and the atmosphere, can be described in terms of the asymmetric fluid theory; we can mention here two important relations: the relation between the stress rates and formation of the dynamic viscous dislocations valid for an incompressible fluid (the 2D turbulence structure), and further cases which might lead us even to molecular structures, like the formation of a foam.

SOLIDS. The introduced balance relations for the antisymmetric part of stresses remain equivalent to those for the stress couples. In the former chapters we have introduced an invariant form of the motion equations for the twist field as defined on the basis of deviatoric strains and presenting the off-diagonal shear oscillations. The full system of motion equations contains the wave forms for the axial strains, spin, and twist.

Now, we return to fracture processes; the fracture phenomena appear with simultaneous changes of material properties including the granulation processes and material crushing. Thus, the constitutive laws and the structural indexes undergo considerable changes; finally, we arrive at a narrow mylonite zone adjacent to fracture. The stresses and deformations may be partly neglected while their time-rates appear important instead. For the deviatoric fields and rotational ones we may write the constitutive laws as follows:

$$S^D_{(ik)} + \tau \dot{S}^D_{(ik)} = 2\mu E^D_{(ik)} + 2\eta \dot{E}^D_{(ik)}, \qquad S_{[ik]} + \tau \dot{S}_{[ik]} = 2\mu E_{[ik]} + 2\eta \dot{E}_{[ik]} \tag{11.22}$$

A mylonite zone might be partly melted.

The progressive fracture with the dynamic angular deformations and oscillation of the off-diagonal shear axes leads to the bond breaking processes, and a released rebound spin arises; the bond breaking process precedes a rebound spin and slip movement, while a rebound spin motion would be retarded in phase.

The presented mechanism may be supported by the special solution for the twist (shear strain) and spin (rotation strain) waves in a mylonite material; however, first we should remind that in a special coordinate system the strain can be presented as the special vectors $\breve{E}_{[s]}$ and $\hat{E}_{(s)}$ (see relations in Chap. 12) and we may assume their common interrelation defining the mylonite state:

$$\breve{E}_{[s]} = \mathrm{i}\hat{E}_{(s)}; \quad \breve{E}_{[s]} = \overset{0}{\breve{E}}_{[s]} exp[\mathrm{i}(k_i x_i - \varpi t)], \quad \hat{E}_{(s)} = \hat{E}^0_{(s)} exp[\mathrm{i}(k_i x_i - \varpi t)] \tag{11.23a}$$

For such a case we may define the complex solutions with the six homogeneous conditions to be fulfilled also in the systems $\{r, \varphi, z\}$:

$$\breve{E}_{[r]} = \mathrm{i}\hat{E}_{(r)}, \quad \breve{E}_{[\varphi]} = \mathrm{i}\hat{E}_{(\varphi)}, \quad \breve{E}_{[z]} = \mathrm{i}\hat{E}_{(z)} \tag{11.23b}$$

corresponding to a turbulence structure.

The conjugate solutions (11.23a, b) suggest that the spin rebound motion will be delayed in phase by $\pi/2$; we may expect a correlation between the recorded twist motions and spin motions shifted by $\pi/2$ in phase in some wavelets.

Further, these conjugate solutions can play an important synchronization role in a progressive fracture process; a wave emission from a micro-fracture source due to an influence of the propagating waves could synchronize the neighboring micro-fractures and in this way the same sense of rotations in a whole process can be assured.

Under a compression load, such a synchronization will lead to the same sense of the induced twist-shear and spin motions, while under shear load—to the formation of a long fracture with attenuated slip motions along the perpendicular fragments with the opposite spins.

In this chapter we have presented the asymmetric continuum theory including different types of material states: from the fluid state up to elastic continuum and the granulated/crushed material.

References

Beirão da Veiga H (2007) Vorticity and regularity for viscous incompressible flows under the dirichlet boundary condition. Results and related open problems. J Math Fluid Mech 9:506–516

Teisseyre R (2008a) Introduction to asymmetric continuum: Dislocations in solids and extreme phenomena in fluids. Acta Geophys 56:259–269

Teisseyre R (2008b) Asymmetric continuum: standard theory. In: Teisseyre R, Nagahama H, Majewski E (eds) Physics of asymmetric continua : extreme and fracture processes. Springer, Heidelberg, pp 95–109

Teisseyre R and Górski M (2009a) Fundamental deformations in asymmetric continuum: Motions and fracturing. Bull Seismol Soc Am 99(2B):1028–1039

Teisseyre R, Górski M (2009b) Transport in fracture processes: fragmentation and slip. Acta Geophys 57(3):583–599

Chapter 12
Release-Rebound Processes and Motions

12.1 Introduction

Some important tensor relations, e.g., those related to the shear deviatoric strains and to rotation strains, may be presented as vector relations in a special coordinate system, e.g., in the diagonal or off-diagonal systems. However, these fields can be presented in the 4D invariant forms by means of the invariant Dirac tensors. Thus, we present the 4D relativistic relations for the invariantly presented deviatoric strain and rotation strain vectors in solids and the molecular strain and rotation vectors in fluids. We end this chapter with outlining a suggestion of K. P. Teisseyre (2012, private communication) that the molecular vorticity may be related to a gradient of pressure in the P-wave propagation; to this end we define the molecular vorticity in solids as based on the molecular transport.

12.2 Release-Rebound Processes and Motions

In our considerations some important relations for vectors and tensors were achieved only in a special coordinate system, e.g., in a diagonal or off-diagonal one; however, we demand that any true relation should be presented in an invariant form. In this chapter we argue that these important relations, which may be achieved in the 3D space only in a special coordinate system, become invariant when presented in the 4D space–time system. To this kind of relations belong the release-rebound motions and processes; some of them become very similar to the Maxwell relations for electro-magnetic fields; we may call these relations the 4D Maxwell-like invariant relations.

However, we may first define the rotation strain vector and the molecular strain vector, as used in solids and in fluids:

$$E_{[i]} = \left\{E_{[23]}, E_{[31]}, E_{[12]}\right\}, \quad \text{or} \quad \tilde{E}_{[i]} = \left\{\tilde{E}_{[23]}, \tilde{E}_{[31]}, \tilde{E}_{[12]}\right\} \tag{12.1a}$$

R. Teisseyre and M. Teisseyre-Jeleńska, *Asymmetric Continuum*,
GeoPlanet: Earth and Planetary Sciences, DOI: 10.1007/978-3-642-31860-3_12,

For the shear strains, a similar attempt could be presented only when we use the off-diagonal coordinate systems, $E_{(11)} = E_{(22)} = E_{(33)} = 0$, or $\tilde{E}_{(11)} = \tilde{E}_{(22)} = \tilde{E}_{(33)} = 0$; in such a case we may define the shear strain vectors:

$$E_{(i)} = \left\{E_{(23)}, E_{(31)}, E_{(12)}\right\}, \quad \text{and} \quad \tilde{E}_{(i)} = \left\{\tilde{E}_{(23)}, \tilde{E}_{(31)}, \tilde{E}_{(12)}\right\} \tag{12.1b}$$

The spin vector, $E_{[i]}$ or $\tilde{E}_{[i]}$, describes the rotational oscillations of the off-diagonal shear axes, and the twist vector (shear strain and molecular shear strain related), $E_{(i)}$, or $\tilde{E}_{(i)}$; these fields relate to the changes of an internal shear field. The shear-twist vector, $E_{(s)}$, may represent the off-diagonal oscillations of shear axes and their amplitudes; such oscillations could be excited by internal fracture processes in the seismic sources due to the consecutive micro-fracture energy releases. Similarly to these fields appearing due to the defects and fracture processes, we may consider the molecular shear fields in fluids.

Now we can note that these rotation and shear motions may be defined as the complex vectors related to the rotation strains and shear strains in the off-diagonal system (cf., Chap. 2):

$$E_{[s]} = \frac{1}{2}\varepsilon_{skl} E_{[kl]}, \; \tilde{E}_{[s]} = \frac{1}{2}\varepsilon_{skl} \tilde{E}_{[kl]}; \quad E_{(s)} = \frac{1}{2}|\varepsilon_{skl}| E^{D}_{(kl)}, \; \tilde{E}_{(s)} = \frac{1}{2}|\varepsilon_{skl}| \tilde{E}^{D}_{(kl)} \tag{12.2a}$$

and for the complex vectors:

$$E_s = E_{[s]} + iE_{(s)}, \quad \tilde{E}_s = \tilde{E}_{[s]} + i\tilde{E}_{(s)} \tag{12.2b}$$

Next, we present the basic 4D invariant definitions for the Dirac tensors, originally given in the system $\{\mathbf{x}, \mathrm{i}ct\}$. Similar definitions will be valid also when we will introduce the invariant light velocity, c, divided by certain material constants, κ or $\tilde{\kappa}$, which permits to reduce this velocity according to the given conditions. The invariant Dirac tensors are defined as follows:

$$\gamma^1 = \begin{bmatrix} 0 & 0 & 0 & -1 \\ 0 & 0 & 1 & 0 \\ 0 & 1 & 0 & 0 \\ -1 & 0 & 0 & 0 \end{bmatrix}, \quad \gamma^2 = \begin{bmatrix} 0 & 0 & -1 & 0 \\ 0 & 0 & 0 & -1 \\ -1 & 0 & 0 & 0 \\ 0 & -1 & 0 & 0 \end{bmatrix},$$
$$\gamma^3 = \mathrm{i}\begin{bmatrix} 0 & 0 & 0 & 1 \\ 0 & 0 & 1 & 0 \\ 0 & -1 & 0 & 0 \\ -1 & 0 & 0 & 0 \end{bmatrix}, \quad \gamma^4 = \mathrm{i}\begin{bmatrix} 1 & 0 & 0 & 0 \\ 0 & 1 & 0 & 0 \\ 0 & 0 & -1 & 0 \\ 0 & 0 & 0 & -1 \end{bmatrix}, \tag{12.3a}$$

and

$$\gamma^1\gamma^3 = \mathrm{i}\begin{bmatrix} 1 & 0 & 0 & 0 \\ 0 & -1 & 0 & 0 \\ 0 & 0 & 1 & 0 \\ 0 & 0 & 0 & -1 \end{bmatrix}, \quad \gamma^2\gamma^3 = \mathrm{i}\begin{bmatrix} 0 & 1 & 0 & 0 \\ 1 & 0 & 0 & 0 \\ 0 & 0 & 0 & -1 \\ 0 & 0 & -1 & 0 \end{bmatrix}, \quad \gamma^1\gamma^2 = \begin{bmatrix} 0 & 1 & 0 & 0 \\ -1 & 0 & 0 & 0 \\ 0 & 0 & 0 & -1 \\ 0 & 0 & 1 & 0 \end{bmatrix}$$

$$\gamma^1\gamma^4 = \mathrm{i}\begin{bmatrix} 0 & 0 & 0 & 1 \\ 0 & 0 & -1 & 0 \\ 0 & 1 & 0 & 0 \\ -1 & 0 & 0 & 0 \end{bmatrix}, \quad \gamma^2\gamma^4 = \mathrm{i}\begin{bmatrix} 0 & 0 & 1 & 0 \\ 0 & 0 & 0 & 1 \\ -1 & 0 & 0 & 0 \\ 0 & -1 & 0 & 0 \end{bmatrix}, \quad \gamma^3\gamma^4 = -\begin{bmatrix} 0 & 0 & 0 & 1 \\ 0 & 0 & 1 & 0 \\ 0 & 1 & 0 & 0 \\ 1 & 0 & 0 & 0 \end{bmatrix}, \tag{12.3b}$$

and

$$\gamma^4\gamma^2\gamma^3 = \begin{bmatrix} 0 & -1 & 0 & 0 \\ -1 & 0 & 0 & 0 \\ 0 & 0 & 0 & -1 \\ 0 & 0 & -1 & 0 \end{bmatrix} \tag{12.3c}$$

Using the definitions (12.3) the considered fields can be presented now as the complex invariant vectors:

$$E_{\lambda\kappa} = E_{[lk]} + iE_{(lk)} = E_{23}\gamma^1 + E_{31}\gamma^2 + E_{12}\gamma^4\gamma^2\gamma^3 = \begin{bmatrix} 0 & -E_3 & -E_2 & -E_1 \\ -E_3 & 0 & E_1 & -E_2 \\ -E_2 & E_1 & 0 & -E_3 \\ -E_1 & -E_2 & -E_3 & 0 \end{bmatrix} \tag{12.4a}$$

or

$$E_\sigma = E_{[s]} + iE_{(s)} = E_1\gamma^1 + E_2\gamma^2 + E_3\gamma^4\gamma^2\gamma^3 = \begin{bmatrix} 0 & -E_3 & -E_2 & -E_1 \\ -E_3 & 0 & E_1 & -E_2 \\ -E_2 & E_1 & 0 & -E_3 \\ -E_1 & -E_2 & -E_3 & 0 \end{bmatrix}, \tag{12.4b}$$

The related wave equations with the above-defined material constant, κ or $\tilde{\kappa}$, can be now written as follows:

$$\begin{aligned} &\sum_n \frac{\partial^2}{\partial x_n^2} E_{\lambda n} - \frac{\kappa^2\partial^2}{c^2\partial t^2} E_{\lambda n} = Y_{\lambda n}\,, \quad \text{or} \quad \Box E_{\lambda\eta} = Y_{\lambda\eta} \\ &\sum_n \frac{\partial^2}{\partial x_n^2} \tilde{E}_{\lambda n} - \frac{\tilde{\kappa}^2\partial^2}{c^2\partial t^2} \tilde{E}_{\lambda n} = \tilde{Y}_{\lambda n}\,, \quad \text{or} \quad \Box \tilde{E}_{\lambda\eta} = \tilde{Y}_{\lambda\eta} \end{aligned} \tag{12.5a}$$

or

$$\sum_n \frac{\partial^2}{\partial x_n^2} E_\sigma - \frac{\kappa^2 \partial^2}{c^2 \partial t^2} E_\sigma = Y_\sigma, \quad \text{or} \quad \Box E_\sigma = Y_\sigma$$
$$\sum_n \frac{\partial^2}{\partial x_n^2} \tilde{E}_\sigma - \frac{\tilde{\kappa}^2 \partial^2}{c^2 \partial t^2} \tilde{E}_\sigma = \tilde{Y}_\sigma, \quad \text{or} \quad \Box \tilde{E}_\sigma = \tilde{Y}_\sigma \tag{12.5b}$$

Here, we have put $\frac{c}{\kappa} = \sqrt{\mu/\rho}$, as it may represent a velocity related to crack processes (cf., Wiederhorn 1967; Rice and Simons 1976; Teisseyre 1980; Teisseyre 1985), or to the molecular relations $\frac{c}{\tilde{\kappa}} = \sqrt{\tilde{\mu}/\rho}$.

We have also introduced here the external forces, which are defined as follows:

$$Y_{\lambda\kappa} = \mathrm{i}Y_{23}\gamma^1 + \mathrm{i}Y_{31}\gamma^2 + Y_{12}\gamma^3 = \begin{bmatrix} 0 & Y_{12} & -Y_{31} & -Y_{23} \\ -Y_{12} & 0 & Y_{23} & -Y_{31} \\ Y_{31} & -Y_{23} & 0 & -Y_{12} \\ Y_{23} & Y_{31} & Y_{12} & 0 \end{bmatrix} \tag{12.6a}$$

$$\tilde{Y}_{\lambda\kappa} = \mathrm{i}\tilde{Y}_{23}\gamma^1 + \mathrm{i}\tilde{Y}_{31}\gamma^2 + \tilde{Y}_{12}\gamma^3 = \begin{bmatrix} 0 & \tilde{Y}_{12} & -\tilde{Y}_{31} & -\tilde{Y}_{23} \\ -\tilde{Y}_{12} & 0 & \tilde{Y}_{23} & -\tilde{Y}_{31} \\ \tilde{Y}_{31} & -\tilde{Y}_{23} & 0 & -\tilde{Y}_{12} \\ \tilde{Y}_{23} & \tilde{Y}_{31} & \tilde{Y}_{12} & 0 \end{bmatrix} \tag{12.6b}$$

and quite similarly for the vector forms: Y_σ, $\tilde{Y}_\sigma$.

These complex relations can be split into the following Maxwell-like equations in 3D for the vectors $E_{(s)}$, $E_{[i]}$, defined in Eq. (12.2a, b) and for their molecular forms, $\tilde{E}_{(s)}$, $\tilde{E}_{[i]}$:

$$\varepsilon_{spq} \frac{\partial}{p} E_{[q]} - \frac{1}{V}\frac{\partial}{\partial t} E_{(s)} = \frac{4\pi}{V} J_s; \quad \varepsilon_{spq} \frac{\partial}{\partial x_p} E_{(q)} + \frac{1}{V}\frac{\partial}{\partial t} E_{[s]} = 0$$
$$\varepsilon_{spq} \frac{\partial}{\partial x_p} \tilde{E}_{[q]} - \frac{1}{\tilde{V}}\frac{\partial}{\partial t} \tilde{E}_{(s)} = \frac{4\pi}{\tilde{V}} \tilde{J}_s; \quad \varepsilon_{spq} \frac{\partial}{\partial x_p} \tilde{E}_{(q)} + \frac{1}{\tilde{V}}\frac{\partial}{\partial t} \tilde{E}_{[s]} = 0 \tag{12.7a}$$

where $\frac{\partial E_{[s]}}{\partial x_s} = 0$, $\frac{\partial E_{(s)}}{\partial x_s} = 4\pi\varepsilon$, and we have introduced the defect related current field, J_k, and propagation velocity, V, under the condition that this velocity shall be transformed according to the relativistic rules for a sum of velocities.

A more general presentation of these solutions for $E_{[q]}$, $E_{(q)}$ and $\tilde{E}_{[q]}, \tilde{E}_{(q)}$, could be given as follows:

$$E_{[q]} = A \exp[i\alpha(kr - \omega t)]; \; E_{(q)} = B \exp[i\beta(kr - \omega t)]$$
$$\text{at} \quad A \exp \alpha = B \exp \beta \tag{12.7b}$$

where the introduced constants α, β could lead to the quite different amplitudes A, B for the shear and rotation waves.

These Maxwell-like relations open one important question related to the introduced wave propagation velocity; to assure their invariance we should define properly the related constant propagation velocity, V or $\tilde{V}$. Such a constant velocity might be defined independently for the different materials and thermodynamical conditions; this is a very important question, and therefore we should relate it to the light velocity:

$$V = \frac{c}{\kappa}, \quad \text{or} \quad \tilde{V} = \frac{c}{\tilde{\kappa}} \tag{12.7c}$$

where κ is an adequatelly big constant depending on a given material and related conditions.

As explained further on, the Maxwell-like relations can describe both the strain wave propagation and also the fracture events occuring in the different conditions and materials. Thus, similarly we have assumed that a fracture propagation starts when an applied load and thermodynamical condition lead to a fluidization along a given path (κ or $\tilde{\kappa}$ is an adequatelly big constant depending on a given material and related fracture conditions).

The related 4D form for these Maxwell-like relations (Eq. 12.7a) could be presented as follows:

$$\frac{\partial}{\partial x^{\kappa}} E_{\lambda\kappa} = \frac{4\pi\kappa}{c} J_{\lambda}; \quad x^{\lambda} = \{x^1, x^2, x^3, x^4\}, \; x^4 = \mathrm{i}\frac{c}{\kappa} t \tag{12.8}$$

At a fracture process, e.g., an earthquake, a wave propagation will be excited. The release-rebound theory (cf., Teisseyre 1985, 2011) states that a release of one field, e.g., break of molecular bonds, described as a release of rotation field, $\partial \breve{E}/\partial t$, causes the rebound-release of another field, e.g., rotation of shears, rot$\hat{E}$, which can be recorded as a slip motion; vice versa, a release of shear field, $\partial \hat{E}/\partial t$, causes the rebound-release of a rotation of the angular strains, rot$\breve{E}$, to be recorded as a rotation field. As we have seen, these release-rebound processes could be very well described by the Maxwell-like relations (Teisseyre 2009).

The related wave mosaic (cf., Chap. 2) explains an interrelated propagation of the shear and rotation motions.

We may note that the motion equations for the symmetric part of strains may lead to the wave equations for the potentials; we may follow the classic definitions:

$$S_{(kl),k} = \rho\,\ddot{u}_l + F_l\,, \quad u_l = l^2 \frac{\partial \varphi}{\partial x_l} + l^2 \varepsilon_{lps} \frac{\partial \psi_s}{\partial x_p}\,, \quad F_l = l^2 \frac{\partial \varphi^F}{\partial x_l} + l^2 \varepsilon_{lps} \frac{\partial \psi_s^F}{\partial x_p} \tag{12.9a}$$

However, instead of the strains system,

$$S_{kl} = S_{(kl)} + S_{[kl]} = \delta_{kl}\bar{S} + \hat{S}_{(kl)} + \breve{S}_{[kl]}; \quad \bar{S} = \frac{1}{3}\sum_{s=1}^{3} S_{(ss)},\ \hat{S}_{(ik)} = S_{(ik)} - \delta_{ik}\frac{1}{3}\sum_{s=1}^{3} S_{(ss)}$$

$$E_{kl} = E_{(kl)} + E_{[kl]} = \delta_{kl}\bar{E} + \hat{E}_{(kl)} + \breve{E}_{[kl]}; \quad \bar{E} = \frac{1}{3}\sum_{s=1}^{3} E_{(ss)},\ \hat{E}_{(ik)} = E_{(ik)} - \delta_{ik}\frac{1}{3}\sum_{s=1}^{3} E_{(ss)}$$

$$S_{(ik)} = 2\mu E_{(ik)} + \lambda\delta_{ik}E_{(ss)} \quad \text{and} \quad \breve{S}_{[ik]} = 2\mu\breve{E}_{[ik]}; \quad \bar{S} = (2\mu + 3\lambda)\bar{E},\ \hat{S}_{(ik)} = 2\mu\hat{E}_{(ik)} \tag{12.9b}$$

we may use the potentials for strains:

$$\begin{aligned} E_{(lq)} &= l^2\left(\frac{\partial^2}{\partial x_l \partial x_q}\Phi + \frac{1}{2}\varepsilon_{lps}\frac{\partial^2}{\partial x_q \partial x_p}\Phi_s + \frac{1}{2}\varepsilon_{qps}\frac{\partial^2}{\partial x_l \partial x_p}\Phi_s\right) \to E_{(lq)}(\Phi) \\ E_{[lq]} &= l^2\left(\frac{1}{2}\varepsilon_{lps}\frac{\partial^2}{\partial x_q \partial x_p}\Psi_s - \frac{1}{2}\varepsilon_{qps}\frac{\partial^2}{\partial x_l \partial x_p}\Psi_s\right) \to E_{[lq]}(\Psi) \end{aligned} \tag{12.10a}$$

where we have put $\frac{\partial\psi_s}{\partial x_s} = 0$, $\frac{\partial\psi_s}{\partial x_s} = 0$ and introduced the intrinsic length unit l.

Now we can present the related wave equations for these potentials:

$$\begin{aligned} \mu\,\Delta E(\Phi)_{(lq)} - \rho\ddot{E}(\Phi)_{(lq)} &= 0 \\ \mu\,\Delta E(\Psi)_{[lq]} - \rho\ddot{E}(\Psi)_{[lq]} &= 0 \end{aligned} \tag{12.10b}$$

And, moreover, we have included here below a sequence of the equal fields: $\frac{\partial^2 u_n}{\partial t \partial x_n} = \frac{\partial v_n}{\partial x_n} = w_n = \frac{\partial D_{nn}}{\partial t} = \frac{\partial E_{nn}}{\partial t} = \tilde{E}_{nn}$, i.e., from the second derivative (space, time) of displacement to a derivative of transport velocity (cf., Chap. 3) and to molecular transport (cf., Chap. 3) and to the time derivatives of deformation and axial strain, and finally to molecular axial strain. Thus, the presented fields include the following derivatives: $E_{\lambda\kappa}\frac{\partial E_{\lambda\lambda}}{\partial t} \to \left\{\frac{\partial E_{\lambda\lambda}}{\partial t}\right\} \to \frac{\partial u_\lambda}{\partial x^\lambda \partial t} \leftrightarrow w_\lambda$ with $x^4 = iVt$.

Thus, we have presented the invariant relations for the shear-twist and rotation wave system in 4D. Let us recall a simple propagation mechanism related to the deviatoric strain and rotation fields in 4D (cf. Teisseyre 2009), which has a lot of analogy to the Maxwell electromagnetic relations. According to the derived relations (12.1) the rotation field and the deviatoric strains, at a constant pressure, shall fulfill the wave equations; the equations for the shear deviatoric strains, $E^D_{(ik)}$, may be reduced, in a special off-diagonal coordinate system with $E^D_{(11)} = E^D_{(22)} = E^D_{(33)} = 0$, to the three independent vector-like relations (see Eq. 12.2a, b).

According to the derived relations (12.2) the rotation field and the deviatoric strains, at a constant pressure, shall fulfill the wave equations.

We should only remember our invariant definition of the vector $\tilde{E}$; therefore, when comparing these theoretical relations with some experimental results, we shall transform these experimental data to the off-diagonal shear system.

Compression load might induce defects forming the centers with opposite-sense shears and rotations: the micro-breaks lead to the fragmentation process followed by the rebound slip. The shears create dynamic angular deformations, the bond breaks and slip appear as a rebound phase-retarded process; the induced shears lead to fragmentation and granulation preceding the slip rebounds.

Some material micro-destructions preceding the earthquakes or induced seismic events relate to very complicated nonlinear processes. The appearing micro-fractures may have a quite different character than that related to an external load; this is due to the role of internal defects existing as an integral element of any natural material. Thus, we may better understand the forwarding micro-destruction processes before any seismic event. Due to the existing differences related to the various defect densities, we may explain the different effects caused by a given load. Therefore, we should take into account the different effects than those expected due to the given boundary or initial conditions. Thus, these different effects are related to the unknown distribution of the internal defects and its changes due to human activity.

The theory of dislocation and dislocation arrays (Eshelby et al. 1951, cf., Rybicki 1986) bring us closer to understanding the roles of the defect densities and applied stresses; the defect densities influence an effective strain distribution (cf., Kossecka and DeWitt 1977 and Teisseyre and Górski 2012).

These considerations go deeper into the fracture processes and dynamic synchronization. The fracture phenomena are related to the simultaneous changes of material properties and to the granulation and material crushing. A mylonite zone might be partly melted; therefore, we should consider some changes into the constitutive laws for the deviatoric and moment related loads, $S^D_{(ik)}$, $S_{[ik]}$. These conjugate solutions suggest that the spin rebound motion will be delayed in phase by $\pi/2$; we may expect a correlation between the recorded twist motions and spin motions shifted by $\pi/2$ in phase in some wavelets.

Further, these conjugate solutions can play an important synchronization role in a progressive fracture process; a wave emission from a micro-fracture source due to an influence of propagating waves could synchronize the neighbouring micro-fractures and in this way the same sense of rotations in a whole process can be assured.

Under a compression load, such a synchronization will lead to the same sense of the induced twist-shear and spin motions, while under shear load—to the formation of a long fracture with attenuated slip motions along the perpendicular fragments with the opposite spins.

Finally, we should come to the fracture-related equations based on the release-rebound processes applied to the molecular transport vector (cf., Chap. 3) and to its related rotation molecular vector, $\ddot{\omega}_s$, meaning some sub-molecular rotation: we will consider the related molecular fields (the molecular displacements), $w_s = \frac{\partial v_s}{\partial x_s}$ (cf., Chap. 3, Eq. 3.17), and molecular rotations:

$$\begin{aligned} w_n &\to \left(w_r, w_\varphi, w_z = 0\right) \\ \ddot{\omega}_s &\to \left(\ddot{\omega}_r = 0,\ \ddot{\omega}_\varphi = 0,\ \ddot{\omega}_z\right) \end{aligned} \tag{12.11a}$$

where $\ddot{\omega}_s$ and w_r and w_φ are subjected to release-rebound fracture equations with specific velocity-like term v:

$$\varepsilon_{spq}\frac{\partial}{\partial x_p}\ddot{\omega}_q - \frac{\partial}{v\,\partial t}w_s = 0; \qquad \varepsilon_{spq}\frac{\partial}{\partial x_p}w_q + \frac{\partial}{v\,\partial t}\ddot{\omega}_s = 0 \tag{12.11b}$$

where

$$\begin{aligned}\varepsilon_{spq}\frac{\partial}{\partial x_p}\ddot{\omega}_q &\to \left\{\frac{\partial\ddot{\omega}_z}{r\partial\varphi}, -\frac{\partial\ddot{\omega}_z}{\partial r}, 0\right\} \quad \text{and at} \quad \frac{\partial w_z}{\partial r} = 0\\ \varepsilon_{spq}\frac{\partial}{\partial x_p}w_q &\to \left\{-\frac{\partial w_\varphi}{\partial z}, \frac{\partial w_r}{\partial z}, \frac{\partial(rw_\varphi)}{r\partial r} - \frac{\partial w_r}{r\partial\varphi}\right\}\end{aligned} \tag{12.11c}$$

We obtain three relations for w_r, w_φ, $\ddot{\omega}_z$ and at $\frac{\partial w_\varphi}{\partial z} = \frac{\partial w_r}{\partial z} = 0$:

$$\frac{\partial\ddot{\omega}_z}{r\partial\varphi} - \frac{\partial w_r}{v\partial t} = 0, \quad \frac{\partial\ddot{\omega}_z}{\partial r} + \frac{\partial w_\varphi}{v\partial t} = 0, \quad \frac{\partial w_\varphi}{\partial r} - \frac{\partial\ddot{\omega}_z}{v\partial t} = 0 \tag{12.12a}$$

or

$$\frac{\partial}{\partial t}\frac{\partial\ddot{\omega}_z}{r\partial\varphi} - \frac{\partial^2 w_r}{v\partial t^2} = 0, \quad \frac{\partial}{\partial t}\frac{\partial\ddot{\omega}_z}{\partial r} + \frac{\partial^2 w_\varphi}{v\partial t^2} = 0, \quad \frac{\partial}{\partial t}\frac{\partial w_\varphi}{\partial r} - \frac{\partial^2\ddot{\omega}_z}{v\partial t^2} = 0 \tag{12.12b}$$

We have obtained finally a quite new system of relations for w_r, w_φ, $\ddot{\omega}_z$, with a special velocity-like term v; eliminating from relations (12.12a) and (12.12b) the terms related to molecular rotations we obtain for the molecular displacements:

$$\begin{aligned}\frac{\partial^2 w_\varphi}{r\partial\varphi\partial r} - \frac{1}{v^2}\frac{\partial^2 w_r}{\partial t^2} &= 0\\ \frac{\partial^2 w_\varphi}{\partial r^2} + \frac{1}{v^2}\frac{\partial^2 w_\varphi}{\partial t^2} &= 0\end{aligned} \tag{12.12c}$$

and reversely the molecular rotation:

$$\frac{\partial^2\ddot{\omega}_z}{\partial r^2} + \frac{1}{v^2}\frac{\partial^2\ddot{\omega}_z}{\partial t^2} = 0 \tag{12.12d}$$

The last relations present the fracture equations under the release-rebound processes related to the molecular fields.

In the frame of the new theory we have defined the dislocation densities in an equivalent form to the classic definition established by Kossecka and DeWitt (1977). This new definition permits to derive the relations between the asymmetric stresses and dislocation densities in different material states.

The asymmetric continuum theory includes a transition to the states close to fracture processes and further to the fluid state. Considering the fracture processes, we have introduced the hypothesis on the wave synchronization confirmed by the special complex solution for the spin and twist fields.

According to K. P. Teisseyre (2012, private communication) we might postulate an additional relation, similar to the vector relations for shear strain and rotation strains (e.g., Eq. 12.7) and presenting the second counterpart of an interplay between the rotational motions and strains in the plane of the wavefront (and in the consecutive planes crossing the ray path). However, this wave should propagate with the speed which depends on density variations, that is—with the P-wave velocity.

In the classic approach, the vortex processes (cf., Chaps. 3 and 4) are described using the definition of vorticity, ς, and similarly we can define the molecular vorticity in solids defined as follows:

$$\varsigma_k = \varepsilon_{kpi}\frac{\partial \upsilon_i}{\partial x_p} \rightarrow \Xi_k = \varepsilon_{kpi}\frac{\partial w_i}{\partial x_p} \tag{12.13a}$$

where w_i represents the molecular transport (Eq. 3.17): $w_s = \frac{\partial \upsilon_s}{\partial x_s} \rightarrow \left\{w_1 = \frac{\partial \upsilon_1}{\partial x_1},\right.$ $\left. w_2 = \frac{\partial \upsilon_2}{\partial x_2}, w_3 = \frac{\partial \upsilon_3}{\partial x_3}\right\}$.

Such molecular vorticity we may relate to a gradient of pressure appearing in the P-wave propagation, $\partial p/\partial r$; far from the source, we will use further on the (r, φ, z) system:

$$\varepsilon_{spq}\frac{\partial}{\partial x_p}\Xi_q - \frac{\partial}{V^P\partial t}\frac{\partial p}{\partial x_s} = 0, \quad \text{for } s = r \rightarrow \frac{\partial}{r\partial\varphi}\Xi_z - \frac{\partial}{\partial z}\Xi_\varphi - \frac{\partial}{V^P\partial t}\frac{\partial p}{\partial r} = 0 \tag{12.13b}$$

and the conditions for the molecular transport:

$$\varepsilon_{spq}\frac{\partial}{\partial x_p}\frac{\partial p}{\partial x_q} + \frac{\partial}{V^P\partial t}\Xi_s = 0, \quad \text{for } q = r \text{ and for } \frac{\partial p}{\partial\varphi} = 0 \text{ and for } \frac{\partial p}{\partial z} = 0:$$
$$\rightarrow \quad -\frac{\partial}{\partial t}\frac{\partial w_\varphi}{\partial r} + \frac{\partial}{\partial t}\frac{\partial w_r}{r\partial\varphi} = 0, \quad \frac{\partial}{\partial t}\frac{\partial w_r}{\partial z} - \frac{\partial}{\partial t}\frac{\partial w_z}{\partial r} = 0 \tag{12.13c}$$

where $\Xi_k = \varepsilon_{kpi}\frac{\partial w_i}{\partial x_p} = \{\Xi_r, \Xi_\varphi, \Xi_z\} = \left\{\frac{\partial w_z}{r\partial\varphi} - \frac{\partial w_\varphi}{\partial z}, \frac{\partial w_r}{\partial z} - \frac{\partial w_z}{\partial r}, \frac{\partial w_\varphi}{r\partial r} - \frac{\partial w_r}{r\partial\varphi}\right\}$

Then, we might write for the pressure radial derivatives along the radial wave propagation:

$$\frac{\partial}{r\partial\varphi}\Xi_z - \frac{\partial}{\partial z}\Xi_\varphi - \frac{\partial}{V^P\partial t}\frac{\partial p}{\partial r} = 0; \quad \varepsilon_{zr\varphi}\frac{\partial}{\partial x_p}\frac{\partial p}{\partial r} + \frac{\partial}{V^P\partial t}\Xi_z = 0$$
$$\text{or} \quad \frac{\partial}{r\partial\varphi}\Xi_z - \frac{\partial}{\partial z}\Xi_\varphi - \frac{\partial}{V^P\partial t}\frac{\partial p}{\partial r} = 0; \quad \frac{\partial^2 p}{\partial r^2} + \frac{\partial}{V^P\partial t}\Xi_z = 0 \tag{12.13d}$$
$$-\frac{\partial}{\partial z}\frac{\partial\Xi_r}{\partial z} - \frac{\partial}{V^P\partial t}\frac{\partial p}{\partial r} = 0; \quad \frac{\partial^2 p}{\partial r^2} + \frac{\partial}{V^P\partial t}\frac{\partial\Xi_\varphi}{r\partial r} = 0$$

In such a way in a solid continuum there might appear the molecular rotations in the pressure propagation waves and related events. In this way it is possible to explain an abundance of rotation motions which occur together and just after the P-wave arrivals.

At the end of our considerations we should underline that the presented system of the shear and rotation wave propagation (e.g., 12.5a) for shear and rotation tensors, or the related shear and rotation vectors (Eqs. 12.5b, 12.7, 12.8) means, of course, the correlations of the related space and time derivatives; we obtain an interaction between these fields with relations combining directly the derivatives of shears and rotations (in tensor or vector forms). However, in these joint systems of wave propagation, the amplitudes of related shear and rotational waves depend on the kind of the fracture process: the shear and rotation amplitudes might be very different, depending on various very different fracture processes. The shear and rotation fracture processes might be very different and the wave fields caused, e.g., by a shear fracture could emit the waves with prominent shear amplitudes and with almost vanishing rotations, but with the prominent derivatives of rotations (cf., Eq. 12.7b).

We can also note that a similar procedure to that applied to relations for strains in solids (Eqs. 12.7, 12.8) might be used for the molecular strains in fluids (cf., Chap. 2, Eqs. 2.20–2.26) and we may present the equations of fluid motions for the molecular strains in an invariant 4D release-rebound system.

References

Eshelby JD, Frank FC, Nabarro FRN (1951) The equilibrium of linear arrays of dislocations. Philos Mag 42:351–364

Kossecka E, DeWitt R (1977) Disclination kinematic. Arch Mech 29:633–651

Rice JR, Simons DA (1976) The stabilization of spreading shear faults by coupled deformation-diffusion effects in the fluid-interfiltred porous materials. J Geophys Res 81:5322–5334

Rybicki K (1986) Dislocations and their geophysical applications. In: Teisseyre R (ed) Continuum theories in solid earth physics. Elsevier-PWN, Amsterdam-Warszawa, pp 18–186

Teisseyre R (1980) Earthquake premonitory sequence—dislocation processes and fracturing. Boll Geofis Teor Appl 22:245–254

Teisseyre R (1985) Creep-flow and earthquake rebound: system of the internal stress evolution. Acta Geophys 33(1):11–23

Teisseyre R (2009) Tutorial on new development in physics of rotation motions. Bull Seismol Soc Am 99(2B):1028–1039

Teisseyre R (2011) Why rotation seismology: confrontation between classic and asymmetric theories. Bull Seismol Soc Am 101(4):1683–1691

Teisseyre R, Górski M (2012) Induced strains and defect continuum theory: internal reorganization of load. Acta Geophys 60(1):24–42

Wiederhorn SM (1967) Influence of water vapor on crack propagation in solid-lime glass. J Am Ceram Soc 50:407–414

Printed by Publishers' Graphics LLC
DBT131120.15.18.74